THE COMMONWEALTH AND INTERNATIONAL LIBRARY
Joint Chairmen of the Honorary Editorial Advisory Board
SIR ROBERT ROBINSON, O.M., F.R.S., LONDON
DEAN ATHELSTAN SPILHAUS, MINNESOTA
Publisher: ROBERT MAXWELL, M.C., M.P.

THERMODYNAMICS AND FLUID MECHANICS DIVISION
General Editors: J. H. HORLOCK, W. A. WOODS

THERMOMECHANICS:

THE GOVERNING EQUATIONS

The design upon the cover represents the variation of pressure in a fluid under gravity, the deflection of an elastic solid in shear, the distribution of temperature during the application of heat, the variation of velocity in a fluid under uniform shear, the distribution of voltage in a uniform conductor, and the distribution of concentration under steady diffusion. It is left as an exercise to the reader to determine the corresponding physical significances of the spacing of the horizontal lines.

THERMOMECHANICS

*An introduction to the governing equations of thermodynamics
and of the mechanics of fluids*

BY

J.C. GIBBINGS

PERGAMON PRESS
OXFORD · LONDON · EDINBURGH · NEW YORK
TORONTO · SYDNEY · PARIS · BRAUNSCHWEIG

Pergamon Press Ltd., Headington Hill Hall, Oxford
4 & 5 Fitzroy Square, London W.1
Pergamon Press (Scotland) Ltd., 2 & 3 Teviot Place, Edinburgh 1
Pergamon Press Inc., Maxwell House, Fairview Park, Elmsford,
New York 10523
Pergamon of Canada Ltd., 207 Queen's Quay West, Toronto 1
Pergamon Press (Aust.) Pty. Ltd., 19a Boundary Street,
Rushcutters Bay, N.S.W. 2011, Australia
Pergamon Press S.A.R.L., 24 rue des Écoles, Paris 5^e
Vieweg & Sohn GmbH, Burgplatz 1, Braunschweig

Copyright © 1970 Pergamon Press Ltd.

All Rights Reserved. No part of this pnblication may be reproduced- stored in a retrieval system, or transmitted, in any form or by any means, electronic, mechanical, photocopying, recording or otherwise, without the prior permission of Pergamon Press Ltd.

First edition 1970
Library of Congress Catalog Card No. 74–82382

Printed in Hungary

This book is sold subject to the condition
that it shall not, by way of trade, be lent,
resold, hired out, or otherwise disposed
of without the publisher's consent,
in any form of binding or cover
other than that in which
it is published.

08 006333 0 (flexicover)
08 006334 9 (hard cover)

"Human reason is weak, and may be deceived...."

(THOMAS À KEMPIS, *The Imitation of Christ*, Book 4, ch. 18, para. 4)

This book is dedicated to my wife.

CONTENTS

Editorial Introduction   xi

Preface   xiii

Notation   xvii

1. Summary of the Newtonian Mechanics of Rigid Bodies   1

     1.1 Distance. 1.2 Time. 1.3 Velocity. 1.4 Acceleration. 1.5 Force, Mass, and Momentum. 1.6 Weight. 1.7 Work and Power. 1.8 Work reaction. 1.9 Conservative and stationary fields of force. 1.10 Energy. 1.11 Additive nature of work. References.

2. Mechanical Properties of Infinitesimal Elements   14

     2.1 The continuum. 2.2 Density. 2.3 Surface tension. 2.4 Stresses. 2.5 Pressure. References.

3. Temperature and the Zero'th Law   24

     3.1 Thermal equilibrium. 3.2 Zero'th law of thermodynamics. 3.3 The thermometer. 3.4 Temperature. 3.5 Mercury-in-glass thermometer. 3.6 Continuum limitations of temperature. References.

4. Units   33

     4.1 Choice of units. 4.2 Time. 4.3 Length. 4.4 Force and mass. 4.5 Further derived mechanical units. 4.6 Temperature. 4.7 Current. 4.8 Light. References.

5. The System of Finite Size   39

     5.1 The system. 5.2 The stress in a solid. 5.3 Tangential stress in liquids. 5.4 Tangential stress in gases. 5.5 Variation of pressure in a stationary fluid. 5.6 Continuity of stress.

5.7 Temperature. 5.8 State and property changes. 5.9 Classification and derivation of properties. 5.10 Work. 5.11 Evaluation of work. 5.12 Heat. 5.13 Algebraic characteristics of heat. 5.14 Evaluation of heat. 5.15 The equilibrium state. References.

6. The First Law of Thermodynamics 66

6.1 Cyclic processes. 6.2 First Law of Thermodynamics. 6.3 The measure of J. 6.4 Internal energy. 6.5 Work done by body forces. 6.6 Application of electricity. References.

7. The Manner of Heat Processes 79

7.1 The conduction of heat. 7.2 Temperature distribution along a bar. 7.3 The radiation of heat through a vacuum. 7.4 The radiation of heat through an opaque medium. 7.5 The rate of a process. 7.6 The directional nature of heat by conduction. References.

8. Application of the First Law of Thermodynamics to Solids 100

8.1 The sliding friction process. 8.2 The constant temperature stressing process. 8.3 The effect of hydrostatic pressure. 8.4 Work done by gravity. 8.5 The heat process. 8.6 Equations of state. 8.7 The two property substance. 8.8 The internal energy. 8.9 Internal energy of an elastic metal. 8.10 Internal energy of rubber. 8.11 Application of the First Law of Thermodynamics to the stretching of steel. 8.12 Application of the First Law of Thermodynamics to the stretching of rubber. 8.13 The rate equation. References.

9. The State of Motionless Fluids 128

9.1 The equation of state of liquids. 9.2 The equation of state of gases. 9.3 The gas thermometer. 9.4 Partial pressures. 9.5 Pressure distribution in a liquid under gravity. 9.6 Forces on surfaces immersed in liquids. 9.7 Pressure distribution in a gas under gravity. 9.8 The internal energy of a liquid. 9.9 The internal energy of a gas. 9.10 Components of a pressure force. 9.11 Pressure forces on fluid volumes. References.

10. Mixtures of Phases   161

10.1 Phase distinction. 10.2 The liquid–gas boundary. 10.3 Phase changes. 10.4 Phase boundaries. 10.5 Triple point. 10.6 Critical point. References.

11. The Characteristics of Fluid Motion   173

11.1 Streamlines. 11.2 Pathlines. 11.3 Steady and unsteady flow. 11.4 Fixed and moving axes. 11.5 Two-dimensional flow. 11.6 Shear stress in a moving fluid. 11.7 Comparison of normal and tangential stresses. 11.8 Acceleration effects. 11.9 Pressure in a moving fluid. 11.10 Rotation in a moving fluid. 11.11 The vortex. 11.12 Turbulence. 11.13 Boundary layer and wake flow. 11.14 Compressible and incompressible flow. References.

12. Conservation of Mass in a Fluid Flow   204

12.1 Flow through a streamtube. 12.2 Flow through a control volume. 12.3 Zero time derivative. 12.4 Mean velocity. 12.5 Application of the continuity equation. 12.6 Diffusion. 12.7 Conservation of numbers of particles. 12.8 Conservation of mass. 12.9 Coefficients of diffusion. 12.10 The semipermeable membrane. 12.11 Self-diffusion. References.

13. The Equations relating Process Phenomena   227

13.1 Summary of the basic equations. 13.2 Interaction between phenomena. References.

14. The Momentum Equation for Fluids in Motion   231

14.1 Flow of an element of fluid. 14.2 Relative values of the terms in the Bernoulli equation. 14.3 Variation of properties normal to streamlines. 14.4 Viscous flow between plates. 14.5 Flow through a control volume. 14.6 Physical significance of the wake. 14.7 Unsteadiness in the flow past an aerofoil. 14.8 Angular momentum for a control volume. 14.9 Pumping by a ducted fan. 14.10 Momentum equation for the flow with diffusion. References.

15. Application of the First Law of Thermodynamics to Fluids in Motion   257

15.1 Characteristics of the infinitesimal element. 15.2 Work done by pressures on a moving element of fluid. 15.3 Work done on a moving fluid element by gravity forces. 15.4 Work done by shear stresses. 15.5 Heat applied to the element. 15.6 The energy change of an element. 15.7 The energy equation for steady flow along a streamline. 15.8 Enthalpy. 15.9 The energy equation for the flow through a control volume. 15.10 The rate equation for a control volume. 15.11 The energy relation for flow with diffusion. References.

16. The Adiabatic Flow   281

16.1 The non-viscous flow along a streamline. 16.2 Stagnation properties. 16.3 The compressible flow Bernoulli equation. 16.4 Flow with stationary boundaries. 16.5 Volume change of a gas. 16.6 The energy equation for fluid machines. 16.7 The use of mean quantities. 16.8 Inadequacy of the governing equations. References.

INDEX   295

EDITORIAL INTRODUCTION

THE books in the Thermodynamics and Fluid Mechanics division of the Commonwealth Library have been planned as a series. They cover those subjects in thermodynamics and fluid mechanics that are normally taught to mechanical engineering students in a three-year undergraduate course.

Although there will be some cross-reference to other books in the division, each volume will be self-contained. Lecturers will therefore be able to recommend to their students a volume covering the particular course which they are teaching. A student will be able to purchase a short, low-price, soft-cover book containing material which is relevant to his immediate needs, rather than a large volume in which most of the contents are outside his current field of study.

The book meets the immediate requirements of the mechanical engineering student in his undergraduate course, and of other engineering students taking courses in thermodynamics and fluid mechanics.

PREFACE

THOSE who enjoy reading prefaces might, with profit, turn to ones by Professor Partington[†] and to one by Professor Leacock.[‡] But for the benefit of reviewers who, like Sydney Smith "...never read a book before reviewing it, it prejudices a man so much",[§] I ought to explain the purpose of this volume.

Engineering, as a subject, was ordered in two parts on the formation of the Institution of Civil Engineers of London to distinguish its members from military engineers. A charming story, no less likely because it is not fully documented, has it that George Stephenson's wrath was aroused when it was made clear that his membership was not to be sought by this Institution but rather that a written test was required of this northerner as of anybody else.[||] Shortly after that the Institution of Mechanical Engineers was formed in Birmingham with Stephenson as its first president. Divisions multiplied and universities here and abroad accepted them when they condescended to include engineering in their studies.

I think some of these divisions are positively harmful: we now get field theory taught under the heading of electrostatics for electronic engineers, aerodynamics for aeronautical engineers, diffusion for chemical engineers, electromagnetics for

[†] Partington, J. R., *An Advanced Treatise on Physical Chemistry*, Longmans, London, Vol. 1, 1962; Vol. 3, 1957.

[‡] Leacock, S., *Sunshine Sketches of a Little Town*, Bodley Head, London, 1933.

[§] Quoted by J. B. Morton, *Daily Express*, No. 20189, p. 10, 30 April 1965.

[||] Smiles, S., *The Life of George Stephenson, Railway Engineer*, Murray, London, 3rd edn. (rev.), ch. 36, p. 506, 1857.

electrical engineers, hydraulics for civil engineers, thermal conduction for mechanical engineers, and elasticity for structural engineers.

And those who have tried to teach both thermodynamics and the mechanics of fluids in isolation from each other realize the consequent handicap of this separation. So marked is this isolation that a writer has recently ignored all the extent of the mechanics of fluids in claiming the study of "non-equilibrium thermodynamics" as a recent development of just a few years' standing.

Thermodynamics as so often taught is really thermostatics in that it considers states of equilibrium at each end of a process. The engineer needs to enhance the performance of a system that he has designed and so he has to study also the process itself; this latter has largely been the province of the mechanics of fluids and of the elasticity and plasticity of solids. Here I have tried to introduce these three subjects coherently.

But I am not sure that I started with a clean slate. Too often I found myself isolating topics in accordance with traditions which remained rooted at the back of my mind. However, I felt able to remove some favourites hallowed by age; two such are metacentric height and an excess of charts of the properties of steam. I was sad to see them go; the former because it is so very easy to impress the student with a demonstration of good agreement between analysis and experiment, whilst removal of the latter required the preparation of enough new material to fill at least two lectures. The present contents have also been influenced by the great developments in the mechanics of fluids due to the growth of aeronautics. For example, fluids in motion are generalized from the start as being compressible, and the limits of the continuum concept are marked.

There is a difference between thinking and memorizing, between originating and practising, and between understanding and learning, that makes the difference between education and training. The former of each of these is the concern of a university and this distinction has guided the exposition of this volume as an aid to education.

Unfortunately, with rare but heartening exceptions, the student is not primarily interested in understanding. He wishes to pass his university examination which tests the capacity of his memory and his diligence at the practice of speedy solution of standardized problems.

The rare exceptions might heed the following insufficiencies of this volume.

One such lack is referred to in the last section of this book. It is the absence of a discussion of the second law of thermodynamics. This topic is the subject of Dr. Montgomery's book in the same series.[†] Two others are the study of dimensional analysis and the experience of experimentation. These are discussed in the same series by Mr. Bradshaw.[‡]

Thermomechanics has been built upon the results of experimental observation. As Saint-Exupéry made one of his characters say: ' "Experience will guide us to the rules", he said. "You cannot make rules precede practical experience." '[§] This is almost universally true. A purpose of analysis is to extend the rules derived from experimental observation to cases for which experiment has not been performed. But analysis uses a model of the real event, and so the approximation of the model must be fully appreciated. Sometimes, as with turbulence, a model is so complex as to prohibit comprehensive analysis; then resort must be made to experiment and to dimensional analysis. Also, use by the reader of methods of visualizing the flows described here greatly aids understanding. So the student must be practised in the experimental method. No book suffices.

[†] S. R. Montgomery, *Second Law of Thermodynamics*, Pergamon, 1966.
[‡] P. Bradshaw, *Experimental Fluid Mechanics*, Pergamon, 1964.
[§] Antoine de Saint-Exupéry, *Night Flight*, Allen Lane, ch. 11 (tr. S. Gilbert), 1940.

The subject-matter of this volume comes under the purview of mechanical engineering; but I intended this book to be useful also to, amongst others, students of civil, electrical, aeronautical, and chemical engineering.

The writing of this book has been aided by my university environment. I am grateful to my colleagues for the many stimulating discussions with them. Professor J. H. Horlock has by no means been a nominal editor; his detailed comments at all stages of the manuscript have invariably made me think more deeply. The writings of several authors have influenced me. Those who know their works will appreciate my particular indebtedness to the books of H. Lamb, M. W. Zemansky, N. A. V. Piercy, and J. H. Keenan. And now, with the close of a long period of deferment of household duties, I must acknowledge my gratitude to my wife; her encouragement of this work has been a constant source of strength; she has her own subtle tribute within the text.

Liverpool J. C. GIBBINGS

NOTATION

a	Van der Waals' coefficient § 9.2
a_1, a_2	Virial coefficients § 9.2
$a_{l,n,x}$	Accelerations § 1.4
A_m	Unit atomic mass § 9.2
$\mathcal{A}$	Absorption rate § 7.4
b	Van der Waals' coefficient § 9.2
c	Velocity of light *in vacuo* § 7.3
	Molar concentration § 12.6
c_e	Extinction coefficient § 7.4
$\overline{c_m^2}$	Mean square molecular speed § 7.3
C	Particle concentration §§ 7.3, 12.6
	Compressibility of liquid § 9.1
C_f	Friction coefficient § 14.4
C_h (C_p)	Enthalpy coefficient § 15.8
C_u (C_v)	Internal energy coefficient §§ 8.8, 9.8
D	Diffusion coefficient § 12.9
	Drag § 14.6
e	Internal energy per mass unit § 6.4
e_p	Photon energy § 7.3
e_T	Total energy per mass unit § 15.6
E	Internal energy § 6.4
E'	Energy per volume unit § 7.3
E_T	Total energy § 8.13
F	Force § 1.5
F_g	Weight force § 1.6
F_x, F_y, F_z	Components of force § 1.5
F_β	Body force § 6.5
F_{12}, F_{21}	View factors § 7.3
g	Acceleration due to gravity § 1.6
h	Enthalpy per mass unit § 15.8
h_a	Atmospheric equivalent liquid height § 9.6
H	Enthalpy § 15.11
i	Electric current § 3.1
J	Mechanical equivalent of heat § 6.2
k	Coefficient of thermal conductivity § 7.1
	Radius of gyration § 9.6
K_n	Knudsen number § 2.1

NOTATION

l	Distance §1.1
L	Molecular mean free path §2.1
m	Mass §1.5
$\dot{m}$	Mass flow rate §§2.1, 12.1
M	Molecular molar mass §9.2
n	Distance §§7.1, 14.3
N	Number of molecules §9.2
p	Pressure §2.5
p_a	Atmospheric pressure §9.5
p_1, p_2	Partial pressures
P	Total pressure §14.1
P_o	Poisson ratio §8.4
q	Velocity §1.2
	Electrical charge §6.5
$\dot{q}$	Acceleration §2.5
Q	Heat §5.12
$\dot{Q}$	Rate of heat §5.14
Q_{to}, Q_{by}	Heat to, by a system §§6.2, 6.5
$\dot{Q}^{(\alpha)}, \dot{Q}^{(i)}, \dot{Q}^{(e)}$	Radiant heat rate absorbed, incident, emitted §7.3
r	Radius §§1.4, 2.3, 8.4
	Proportion of phase §10.3
	Compression ratio §16.5
r_x	Radial coordinate §1.4
R	Gas constant §9.2
t	Time §1.3
	Celsius temperature §4.6
T	Absolute temperature §3.4
	Kelvin temperature §4.6
T	Tensile force §8.2
u	Internal energy per mass unit of a two-property substance §8.8
U	Internal energy of a two-property substance §8.8
v	Volume per mole §9.2
v^*	Mean concentration velocity §12.6
V	Volume §2.2
w	Width §2.5
W	Work §1.7
$\dot{W}$	Power §1.7
W_{to}, W_{by}	Work to, by, a system §§6.2, 6.5
x	Distance §1.4
	Property per mass unit §5.9
$\dot{x}$	Velocity §1.4
X	Property §5.9
y	Distance §1.4
$\dot{y}$	Velocity §1.4

NOTATION

y'	Distance in liquid below arbitrary origin § 9.6
$\bar{y}$	Distance to centroid § 9.6
y_p	Distance to point of action of hydrostatic force § 9.9
Y	Young's modulus § 8.2
Y_0	Young's modulus at zero extension § 8.6
z	Height § 1.1
$\dot{z}$	Velocity § 1.4
Z	Compressibility factor § 9.2
Z_H	Total head § 14.1
α	Angle § 2.5
	Proportion of radiation absorbed § 7.3
	Coefficient of linear thermal expansion § 8.5
β	Body force per mass unit § 5.11
	Coefficient of volume thermal expansion § 8.5
γ	Ratio of enthalpy to energy coefficients (Ratio of specific heats) §§ 9.7, 16.1
ε	Small quantity § 2.2
	Dielectric coefficient § 6.5
	Ratio of surface emission to black body emission § 7.3
η	Intensity of heat rate § 5.14
	Fan efficiency § 16.6
θ	Angle §§ 2.5, 9.6
θ_x	Angular velocity of radial ordinate § 1.4
$\varkappa$	Electrical conductivity § 3.1
μ	Coefficient of viscosity § 5.3
ν	Kinematic coefficient of viscosity § 11.11
ϱ	Density § 2.2
σ	Normal stress § 2.4
τ	Tangential stress § 2.4
τ_0	Tangential shear stress at wall in a flow § 5.3
υ	Surface tension § 2.3
ϕ	Electrical voltage § 3.1
ω	Angular velocity of a fluid element § 11.6
$\delta, d, D, \Delta, \infty$	Differential, small difference §§ 2.3, 1.1, 11.4, 1.5, 5.11
$\dagger$	Pertaining to a process § 1.7
$\oint$	Pertaining to a cyclic process § 1.9

CHAPTER 1

SUMMARY OF THE NEWTONIAN MECHANICS OF RIGID BODIES

1.1 Distance

Three fixed points, spaced out along a line, are denoted O, A, and B in Fig. 1.1. It is conventional to speak of a distance from O to A of perhaps 5 cm and of a distance from O to B of perhaps 12 cm. Such statements imply two things.

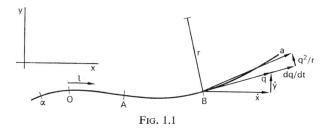

Fig. 1.1

First, distance is not something to which can be ascribed an absolute value for any point. It cannot be said that A has a distance of 5 cm and that B has a distance of 12 cm but only that they have these distances from some origin O. A statement of distances refers to differences and not to absolute values.

Second, before numerical values can be assigned to a difference in position, the path between the two points must be specified: the distance might not be measured along a straight line.

In measuring distance, the origin can usually be chosen quite arbitrarily. This might result in there being points of interest either side of it along the chosen path. No difficulty arises if negative values are assigned to distances measured to one side. For example, the point α, marked in Fig. 1.1 might have a distance from O of -7 cm.

An object travelling along the line would, at any instant, be a certain distance from O. This distance can be thought of as a property of the object. Then before a numerical value can be assigned to it, a value must be chosen for the comparable property at the origin and also the path between the two positions must be defined. This concept is one that will occur again in the present study.

In contrast, a measurement of the difference in vertical height of the object between two points on its path does not require specification of the intervening route. Thus the height, measured as a difference from some arbitrary level,[†] is a single valued function of lattitude and longitude; for any pair of values of the latter the height has only one value. A plot of height on a graph whose ordinates are those of lattitude and longitude is the usual contour map; it is an example of a constant and conservative field of values.

In the rare circumstance of a survey of an area which is gradually subsiding with time, it then becomes necessary to quote a time difference as well as changes in lattitude and longitude when enumerating a height difference. No longer would the field be a constant and conservative one.

A difficulty can arise in the evaluation of a height change. Figure 1.2 illustrates two possible paths along which an object can change its height by an amount $(z_B - z_A)$ between the points marked A and B. Along the first path, denoted I in this figure, each small step of the way results in a smooth change in height

---

[†] Sea-level is one such arbitrary level.

of δz enabling the total change to be determined by the summation

$$z_B - z_A = \int_A^B dz. \tag{1.1}$$

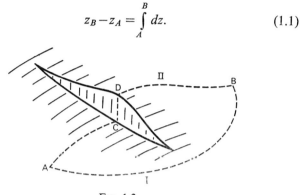

Fig. 1.2

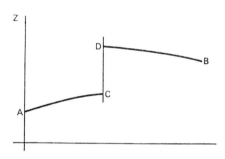

Fig. 1.3

In contrast, the other path II has a discontinuity in height at the vertical precipice CD and now integration, as illustrated in Fig. 1.3, results in

$$\int_A^C dz + \int_D^B dz = z_C - z_A + z_B - z_D$$
$$= (z_B - z_A) - (z_D - z_C)$$

so that computation of the change in height is in error by the amount of the second term.

Along path I, *dz* has the characteristics of a perfect differential. Summation of a perfect differential, as in eqn. (1.1), gives a result that is independent of the path and hence leads to a stationary and conservative field. Study of the changes along path II shows that the converse is not necessarily true; a quantity forming a stationary and conservative field may not be expressible as a perfect differential at all points.

1.2 Time

Time is also measured as a difference from some arbitrary origin such as midnight or midday. However, in the present study there is no question of a path having to be defined, and also an increment of time in a real process is always taken as positive.[†]

1.3 Velocity

Referring again to Fig. 1.1, and denoting by l the distance of an object from the origin and by t the corresponding time, the velocity q is defined as

$$q \equiv \frac{dl}{dt}. \qquad (1.2)$$

The sign of q then corresponds to that of dl.

Velocity being a gradient, then its value is independent of the choice of origins from which both l and t are measured.

The direction of q is taken as the direction of dl, and this, in the limit, is tangential to the path.

---

† To enlarge upon this would lead one into the realms of theology and philosophy; and this is beyond the scope of the present volume.

1.4 Acceleration

The acceleration of an object a in general has components that are tangential and normal to the path (Fig. 1.1). They are given, respectively, by

$$a_l = \frac{dq}{dt} = \frac{d^2l}{dt^2} \tag{1.3}$$

and
$$a_n = q^2/r, \tag{1.4}$$

where r is the radius of curvature of the particle path at the point under consideration and a_n is directed inwards towards the centre of curvature. Alternatively, the x component of the acceleration is obtainable as

$$a_x = \frac{d\dot{x}}{dt} = \frac{d^2x}{dt^2}, \tag{1.5}$$

$\dot{x}$ denoting the corresponding component of velocity. Similar expressions exist for the y and z components.

Another formulation is in terms of polar coordinates. If r_x denotes the radial coordinate of the object from the Ox axis in a plane $x = $ constant, and θ_x denotes the angular velocity of this radius, then the acceleration perpendicular to r_x is[1]†

$$\frac{1}{r_x} \frac{d(r_x^2 \theta_x)}{dt}. \tag{1.6}$$

A change of origin in space, in time, and in velocity, is seen to leave the acceleration unaltered and, again, accelerations are always relative to the acceleration of some arbitrary origin.

---

† Sources referred to in this book are listed at the end of the corresponding chapter.

1.5 Force, mass, and momentum

It has been said that a force is something that can readily be observed through a sense of touch. In fact this sense is a surer guide to stress, that is to force per unit area. Using the same force to press a finger upon the plane surface of a table and then against its sharp corner gives two different sensations.

A force is that which can induce motion in an object. Like acceleration, it is a quantity requiring both a value and a direction to specify it. If the total force applied to a particular object is changed, then there results a change in that object's acceleration. If these changes are measured three things are observed. They are:

(a) The acceleration increment is in the same direction as the force increment.
(b) The force increment ΔF is related to the associated acceleration increment Δa by

$$\Delta F/\Delta a = \text{constant.} \quad (1.7)$$

(c) For a particular object the constant of eqn. (1.7) has a fixed positive value that cannot be changed.

This constant is given a name; it is called the mass of the object m. Though a force and an acceleration can only be measured as differences from arbitrary origins, the mass of an object is an absolute quantity.

Equation (1.7) is a formulation of Newton's law of motion. It being understood that only increments of force and acceleration are measurable, this equation is usually written in the form

$$F = ma. \quad (1.8)$$

Because of (a) above, the component of F in the x direction

F_x is given by

$$F_x = ma_x \qquad (1.9)$$

with similar expressions for other components.

As eqn. (1.7) applies to the motion of an object of constant mass, then eqn. (1.9) can be written

$$F_x = \frac{d}{dt}(m\dot{x}), \qquad (1.10)$$

and it is to be noted that eqn. (1.10) applies also to the motion of an object of time variable mass.[2]† The product $m\dot{x}$ is called the momentum.

1.6 Weight

A special case of eqn. (1.7) is

$$F_g = mg, \qquad (1.11)$$

where F_g is the gravitational acceleration force, called the weight, and g is the resulting acceleration.

1.7 Work and power

A force F being applied to a solid object, and the point of application moving an amount Δl in a direction parallel to the direction of that force, the work done to the object W is defined by

$$W \equiv F \Delta l. \qquad (1.12)$$

The work can have positive or negative values. When the displacement is in the same direction as the force, convention usually assigns a positive value to W.

† See also the discussion of § 14.5.

Work is something that happens whilst motion is in progress. That is why eqn. (1.12) is not written in the form $dW = Fdl$ because this induces an integration between limits resulting in $W_B - W_A = \int_A^B Fdl$. In this expression the left-hand side is meaningless. An object does not contain a certain amount of work W_A at the beginning of a process nor an amount W_B at the end. Here the result of this integration is expressed as

$$\left|_A^B W = \int_A^B F\,dl. \right. \tag{1.13}$$

Particular care must be taken in evaluating this integral. To illustrate why this should be so, consider the movement of a force along a straight line in its own direction for which F is a single valued function of l as sketched in Fig. 1.4. As illus-

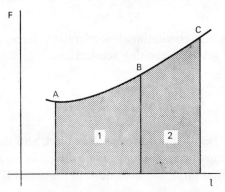

Fig. 1.4

trated, the work done in moving directly from A to B is equal to the area 1, as is the work done in moving indirectly from A to B by overshooting to C and then returning to B because then the area 2 is simply added and subtracted. Integration of eqn. (1.13) with insertion of the limits A and B is straightforward. In contrast, when F is not a single valued function of l as sketched in Fig. 1.5, and the path from A to B' is via B to an overshoot to C and back again, the total work done is now given by the sum of areas 1 and 3, only the area 2 being subtracted as the path is retraced from C back to B'. Integration followed by an insertion of limits is no longer straightforward.

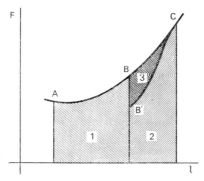

Fig. 1.5

Time increments being always positive in a real process, this difficulty can be avoided by integrating functions of time because these are single valued. Rewriting eqn. (1.13) as

$$\Big|_A^B W = \int_A^B F \frac{dl}{dt}\, dt \tag{1.14}$$

integration is now straightforward.

This integrand is a rate of doing work and is called the power $\dot{W}$ so that, noting eqn. (1.2),

$$\dot{W} \equiv Fq \tag{1.15}$$

and

$$\Big|_A^B W = \int_A^B \dot{W}\, dt. \tag{1.16}$$

1.8 Work reaction

There is an important distinction between the work done by a force applied at the surface of an object and that done by a body force applied within it. With the former there is no discontinuity in velocity between the object and the surroundings applying the force,[†] and as a force has an equal and opposite reaction then the work done to the object is equal to but minus that done by the object to its material surroundings. This is not necessarily so with body forces.

1.9 Conservative and stationary fields of force

In evaluating the work done by gravitational force the value of l is measured relatively to a set of axes. If the gravitational force remains constant with time at any point in this coordinate system, then the work done by it upon an object moving between two points is independent of the path travelled.[3] With reference to Fig. 1.6, the value of $\int_A^B W$ is then the same for path I as for path II as symbolized by

$$\int_A^B W_\mathrm{I} = \int_A^B W_\mathrm{II}.$$

The force at a point being invariable in this case, then also

$$\int_A^B W_\mathrm{I} = -\int_B^A W_\mathrm{I}.$$

An important corollary of these results is that the work done in passing around the circuit from A to B via path I and back

[†] The case of apparent slip is discussed in § 8.1.

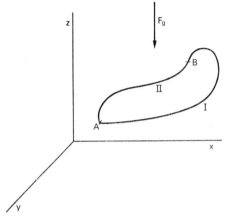

Fig. 1.6

to A by path II has the value

$$\int_A^B W_{\mathrm{I}} + \int_B^A W_{\mathrm{II}} = \int_A^B W_{\mathrm{I}} - \int_A^B W_{\mathrm{II}}$$
$$= \int_A^B W_{\mathrm{I}} - \int_A^B W_{\mathrm{I}}$$
$$= 0,$$

a result expressible as

$$\oint W = 0.$$

Such a process is called a cyclic one. It is important to bear in mind the warning in the previous section which indicates that this result only applies in general for the work done to an object and not for the work done by that object.[†]

Examples of other stationary and conservative force fields for which the cyclic work is zero are:

(a) Electrostatic force fields due to stationary charges.

---

† This is explained in § 6.5.

(b) Magnetic force fields due to stationary permanent magnets of unvarying strength.

(c) Electromagnetic force fields due to invarying direct currents in stationary conductors.

These four types of force are called body forces. Each has its own point of action which can be computed in the manner of obtaining a centre of gravity. Assigning a constant value to g in eqn. (1.11) is an approximation satisfactory for many purposes. Doing so results in the centre of gravity being coincident with the centre of mass. In general, for all these body forces, the point of action does not coincide with the centre of mass.†

1.10 Energy

Writing eqn. (1.9) for the component tangential to the path, multiplying this by q, and substituting from eqn. (1.3), gives

$$\dot{W} = F_l q = mq \frac{Dq}{Dt}$$
$$= \frac{D}{Dt}\left(\frac{1}{2} mq^2\right), \qquad (1.17)$$

where F_l is the component of the force in a direction tangential to the path, and where the D notation implies changes of a single object.‡

A component of the force that is normal to the path will do no work because there is no motion along its line of action.

Integration of eqn. (1.17) then gives

$$\int_A^B W = \tfrac{1}{2} mq_B^2 - \tfrac{1}{2} mq_A^2. \qquad (1.18)$$

† See further discussion in § 11.10.

‡ The significance of this notation becomes apparent in § 11.4.

The quantity $\frac{1}{2}mq^2$ is called the kinetic energy. This equation shows that where a velocity change is directly related to a force by eqn. (1.8), then the change in kinetic energy is equal to the work done.

The kinetic energy, unlike work, can be regarded as something that belongs to the object; it is a property that helps to define the object's state.

1.11 Additive nature of work

The three components F_x, F_y, and F_z of a force F do elements of work given by

$$W_x = F_x\, dx = d(\tfrac{1}{2}m\dot{x}^2) = \tfrac{1}{2}m\, d(\dot{x}^2),$$
$$W_y = F_y\, dy = d(\tfrac{1}{2}m\dot{y}^2) = \tfrac{1}{2}m\, d(\dot{y}^2),$$
$$W_z = F_z\, dz = d(\tfrac{1}{2}m\dot{z}^2) = \tfrac{1}{2}m\, d(\dot{z}^2).$$

Adding these components of work gives

$$\begin{aligned} W_x + W_y + W_z &= \tfrac{1}{2}m[d(\dot{x}^2) + d(\dot{y}^2) + d(\dot{z}^2)] \\ &= \tfrac{1}{2}m\, d(\dot{x}^2 + \dot{y}^2 + \dot{z}^2) \\ &= \tfrac{1}{2}m\, d(q^2). \end{aligned}$$

But the right-hand side of this equation is equal to the work W done by the resultant force. Thus

$$W = W_x + W_y + W_z,$$

and so the total work is obtained by the simple addition of its components.

References

1. LAMB, H., *Dynamics*, Cambridge, 1951, art. 86.
2. BOTTACCINI, M., An alternate interpretation of Newton's second law, *Am. Inst. Aeronaut. Astronaut. J.* **1** (4), 927 (April 1963); **2** (6), 1164 (June 1964).
3. PARS, L. A., *Introduction to Dynamics*, Cambridge, 1953, art. 23.9.

CHAPTER 2

MECHANICAL PROPERTIES OF INFINITESIMAL ELEMENTS

2.1 The continuum

The indication of our senses is that liquids and solids have a continuous structure whereas in reality they have a discontinuous molecular and atomic one; so also have gases.

For the present purposes, solids, liquids, and gases are assumed to have this continuous structure so that their nature is independent of the size of a sample. To satisfy this continuum assumption, the smallest piece examined must be large compared to the spacing of its molecules. In fact it can need to be much larger, for in gases the average distance between molecules is generally very large compared to their size, and so they travel a long way before hitting one another: when this molecular path has a mean length of 1·0 in. in nitrogen then each cubic inch contains 10^{15} molecules.

It is this mean molecular path that provides the criterion justifying the continuum assumption for gases.

The ratio of the molecular mean free path L to some typical dimension of an object is called the Knudsen number K_n. Its significance in limiting the continuum assumption for a fluid flow is illustrated in Fig. 2.1. Plotted are the results of measurements of the rate of mass flow $\dot{m}$ of three different gases through a circular hole in a plate dividing a container.[1] The Knudsen number was identified with the ratio of the molecular mean

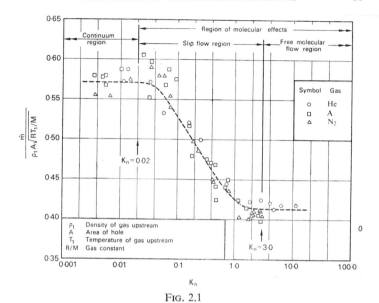

Fig. 2.1

free path in the high pressure side of the container to the hole diameter. It is seen that when K_n has a value below about 0·02 the mass flow rate is independent of the value of the Knudsen number and, furthermore, it is in fair agreement with an approximate solution to the continuum flow which does not consider the detailed molecular effects.[2] A similar limiting value for the Knudsen number has been obtained from experiments upon the thermal equilibrium state of a cylinder in a gas stream.[3] Thus if a flow pattern has a Knudsen number satisfying the approximate relation

$$K_n \leqslant 0.02 \simeq 1/100, \qquad (2.1)$$

existing experiments indicate that the fluid acts as a continuum.

The other two regions, indicated in Fig. 2.1, that correspond to $0.02 < K_n < 3.0$ and to $K_n > 3.0$ are known respectively as the slip flow region and the free molecular flow region. Some authors do not distinguish between these latter two regions, calling them both the microscopic region. The continuum region is then named the macroscopic region.

For liquids, where the mean spacing of the molecules is of order comparable to the molecular size, a significant length for measuring the occurrence of molecular effects has been suggested to be about 100 times this molecular diameter.[4]

2.2 Density

If a quantity of liquid has dimensions that satisfy the inequality of eqn. (2.1), then these dimensions are larger than $10^2 L$ and so the volume size has an order that is larger than $10^6 L^3$. Then the ratio of the mass of liquid M to the volume V is independent of the value of V.† This ratio is called the density ϱ, a definition symbolized by

$$\varrho \equiv \frac{M}{V}, \quad [V \gg 10^6 L^3].$$

In contrast, samples taken from a quantity of gas in motion could have varying values of M/V.

A simple example of the distribution of the mass of a gas along a tube having a constant cross-sectional area A is illustrated in Fig. 2.2. With an x-axis along the tube a mass of gas M is supposed to be contained between the planes $x = 0$ and $x = x$. Its volume V is equal to Ax and, were it a liquid, its density would have the value $M/(Ax)$. With a gas, the distribution of mass is expressed by the series

$$\frac{M}{Ax} = \varrho + \frac{d\varrho}{dx} x + \frac{1}{2} \frac{d^2\varrho}{dx^2} x^2 + \ldots,$$

where $\varrho, \dfrac{d\varrho}{dx}, \dfrac{d^2\varrho}{dx^2}, \ldots$ are constants.

† Equally it must not be too large in a gravitational field (§ 1.9).

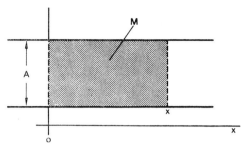

Fig. 2.2

In defining the density at the origin, x can only tend to a finite value that is greater than the molecular mean free path. In later analyses elements are shrunk to a limiting size of zero. For this to be acceptable all the terms in the above series after the first must be negligible, that is,

$$\varrho \gg \frac{d\varrho}{dx} x,$$

$$\varrho \gg \frac{1}{2} \frac{d^2\varrho}{dx^2} x^2,$$

and so on. The first inequality can be expressed as

$$\varepsilon \varrho = \frac{d\varrho}{dx} x,$$

where ε is a small number. Identifying the Knudsen number with L/x, the limiting value of $d\varrho/dx$ for a continuum is

$$\frac{d\varrho}{dx} = \frac{\varepsilon K_n}{L} \varrho.$$

For example, in air, at normal atmospheric conditions, the mean free path L is about $2 \cdot 3 \times 10^{-6}$ in., and so, taking $K_n = 1/100$ and arbitrarily choosing $\varepsilon = 1/100$,

$$\frac{1}{\varrho} \frac{d\varrho}{dx} = 44 \text{ in}^{-1}.$$

This limit is such that the density must not double in less than about one-fortieth of an inch.

A limitation of this order occasionally applies to a fluid flow. It can exist when either a train of very high frequency sound waves or a strong pressure wave[†] is travelling through a gas, and in very low density gas flows.

Adopting these numerical values, the definition of density in a gas in motion is expressed by

$$\varrho \equiv \frac{M}{V}, \quad \left[V \to > 10^6 L^3, \quad \frac{d\varrho}{dx} < (10^{-4}\varrho)/L\right]. \quad (2.2)$$

2.3 Surface tension

At the interface between either a liquid and a gas or between two liquids, the binding force between molecules results, in continuum terms, in a force acting across any imaginary line on this surface. This is illustrated in Fig. 2.3.

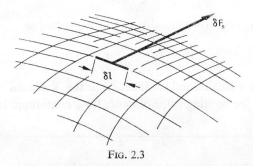

Fig. 2.3

It is found that as an infinitesimal length of the line δl is reduced the following two results are obtained:

† Known as a shock wave.

(a) The direction of the force tends towards the normal to δl and the tangent to the interface surface.
(b) The ratio of the force δF_s to the length of the line δl tends to a constant.[†]

The surface tension v is defined as this ratio but, as with the definition of density, δl is limited to a value that is large compared to the molecular mean free path. Equally, δl must not be too large. On a flat interface the line would be straight and $\delta F_s/\delta l$ would be constant for all values of $\delta l \gg L$. If its value for a curved interface is to approach that for a flat one, then δl must be much smaller than the radius of curvature of the line r. A formal definition of the surface tension then becomes

$$v \equiv \frac{\delta F_s}{\delta l}, \quad [10^2 L < \delta l \ll r]. \tag{2.3}$$

2.4 Stresses

Across a small surface drawn in a substance, molecular interactions result in a continuum force being exerted equally on the material on either side. This is illustrated in Fig. 2.4, where a force δF is shown acting upon the material below an area δa.

This force has the two components δF_n normal to and δF_t tangential to the surface.

A stress is defined as the limit of $\delta F/\delta a$ as δa is reduced in size to a lower limit as before. Hence a normal stress σ is defined by

$$\sigma \equiv \frac{\delta F_n}{\delta a}$$

[†] This statement is amplified later in § 10.2.

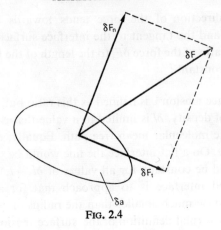

Fig. 2.4

and a tangential stress τ by

$$\tau \equiv \frac{\delta F_t}{\delta a}.$$

These terms are used in the theory of fluids. In the theory of solids τ is called the shear stress whilst σ is called the tensile stress when it has a positive value as drawn in Fig. 2.4, and the compressive stress when it has a negative value.

In both solids and liquids the normal stress can be positive or negative though occurrence of a positive value in liquids is rare.[5] In gases it is negative.

In solids the ratio σ/τ can vary between zero and infinity. In fluids the normal stress is much greater than the tangential stress except on the rarest of occasions.†

2.5 Pressure

Cases of the tangential stress in a fluid being zero are discussed in §§ 5.3 and 11.11. A consequence of this condition is now obtained.

† See also the discussion of § 11.7.

Figure 2.5 is a diagram of a small element of fluid in the form of a prism whose constant cross-sectional shape is a right-angle triangle. A view of this cross-section is drawn in Fig. 2.6 showing one edge to be horizontal.

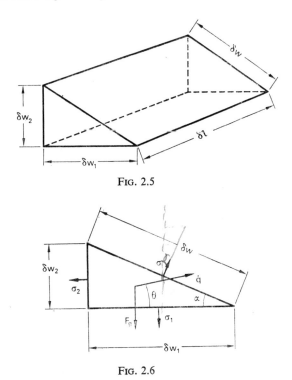

FIG. 2.5

FIG. 2.6

The forces acting upon the element, due to the weight and the normal stresses, result in an accelaration $\dot{q}$ at an angle θ to the horizontal. The horizontal component of the force acting upon the element is

$$\sigma \, \delta w \, \delta l \sin \alpha - \sigma_2 \, \delta w_2 \, \delta l,$$

and so, as this equals the product of the mass and the horizontal component of the acceleration,

$$\sigma\,\delta w\,\delta l\,\sin\alpha - \sigma_2\,\delta w_2\,\delta l = \varrho\,\delta w_1\frac{\delta w_2}{2}\,\delta l\,\dot q\,\cos\theta.$$

This simplifies because

$$\delta w_2 = \delta w\,\sin\alpha$$

to

$$\sigma - \sigma_2 = \tfrac{1}{2}\varrho\dot q\,\cos\theta\,\delta w_1.$$

If the element is now shrunk to zero size, every item in this equation remains finite except δw_1 which also tends to zero. Thus at this limit the right-hand side becomes zero and so

$$\sigma = \sigma_2. \tag{2.4}$$

Similarly, by resolving vertically,

$$\sigma\,\delta w\,\delta l\,\cos\alpha - \sigma_1\,\delta w_1\,\delta l - \varrho g\,\delta w\,\frac{\delta w_2}{2}\,\delta l$$

$$= \varrho\,\delta w_1\frac{\delta w_2}{2}\,\delta l\,\dot q\,\sin\theta,$$

and as

$$\delta w_1 = \delta w\,\cos\alpha,$$

$$\sigma - \sigma_1 = \tfrac{1}{2}\varrho\,\delta w_2(g + \dot q\,\sin\theta).$$

When, as before, the size of the element is shrunk to zero,

$$\sigma = \sigma_1.$$

Combining this result with eqn. (2.4) gives

$$\sigma = \sigma_1 = \sigma_2, \tag{2.5}$$

a result that is independent of the value of α and of the direction of orientation of δl. So, in the absence of tangential stresses, and when body forces are proportional to the mass, the normal stress at a point is the same in all directions. In this special case a pressure p is defined by the relation

$$p \equiv -\sigma. \tag{2.6}$$

References

1. LIEPMANN, H. W., A study of effusive flow, *Aeronautics and Astronautics*, Durand Centennial Conference, Pergamon, Oxford, 1960, p. 153.
2. FANKLE, F. I., *Dokl. Akad. Nauk SSSR (Acad. of Sciences USSR)* **58** (3), 381 (1947).
3. STALDER, J. R., GOODWIN, G., and CREAGER, M. O., *Heat Transfer to Bodies in a High-speed Rarefied Gas Stream*, Nat. Adv. Comm. Aer. Tech. Note 2438, Washington, August 1951.
4. BIKERMAN, J. J., *Surface Chemistry*, 2nd edn., Academic Press, 1958, para. 182, p. 265.
5. REYNOLDS, O., *On the Internal Cohesion of Liquids and the Suspension of a Column of Mercury to a Height more than Double that of the Barometer*, Memoirs, Manchester Lit. Phil. Soc., 3rd Series, 1882, vol. 7, p. 1.

CHAPTER 3

TEMPERATURE AND THE ZERO'TH LAW

3.1 Thermal equilibrium

The sense of touch is a poor guide to levels of temperature, being more nearly indicative of rates of heat being applied to the skin. If a piece of wood and a piece of metal have been together for a while in the same room, then touch indicates a greater rate of heat application to the metal than to the wood even though the characteristic called temperature is the same for both. An indication of temperature that is not sensitive to rates of application of heat is thus required; this is now discussed.

When a poker is placed into a red-hot fire its tip eventually glows with the same shade of the same colour as that of the adjacent part of the fire: some form of equilibrium between the tip of the poker and the adjacent coals is achieved. This type of equilibrium is called thermal equilibrium. Other solid materials which do not burn would similarly glow and can become almost indistinguishable from the surroundings; they can thus act as thermal comparators. For suppose a piece of such material to be transferred so rapidly from one fire to another that in the transfer no change takes place in its colour. And further, suppose that on entering the second fire not only does its colour match the second fire but that is continues to do so. The piece of material of unchanging state would thus be in thermal equilibrium with both furnaces. Were the coals of one furnace to be shovelled into the other with similar rapidity then it would be seen that a thermal equilibrium also existed between the two furnaces; colour comparison with the thermal comparator would have demonstrated the equality of some property of the two furnaces. This property will be called temperature.

Alternatively, this comparison could be made by measurement of the electrical conductivity of a piece of wire.[†] Bringing such a wire into a room from a colder environment would be followed by a continuous change in the conductivity until thermal equilibrium between the wire and the air in the room is attained. Thermal equilibrium between this room and an adjacent one would be indicated by the equilibrium conductivity having the same value in both. Now the wire has acted as a thermal comparator. Or again, suppose that some air, contained in a sealed and rigid container, has been in a place long enough for the air pressure to reach an equilibrium value. On the container being moved to another place, and the pressure being measured again after the attainment of equilibrium, then an equality between these two pressures indicates a thermal equilibrium between the two places.

Similarly, the length of a piece of metal or the volume of a fixed amount of liquid could be used as indicators of thermal equilibrium.

As a sixth example, a metal object could be suspended by a magnetic force and in a vacuum within a container, so that it has no physical contact.[‡] The size of the object could then be

---

[†] The conductivity $\varkappa$ is defined by the relation
$$i = - \operatorname*{Lt}_{i \to 0} \varkappa \frac{\partial \varphi}{\partial n},$$
where i is the current and $\partial \varphi / \partial n$ the gradient of voltage. The reason for the limitation to zero current in this definition should become apparent from the later discussion in §§ 6.6 and 13.1.

[‡] In so far as no realizable vacuum is perfect.

observed by optical means to reach an equilibrium value. On the object being brought into physical contact with the container, its size would remain unchanged so that this thermal equilibrium remains. Transfer to the interior of another vacuum container would enable thermal equilibrium between the two containers to be verified by an observation that the metal object retains the same size.

In each of these methods of confirming thermal equilibrium there must be no change in the state of the thermal comparator and of the objects of comparison during the process of comparison. Additionally, for the first five examples there must be physical contact with no relative motion and with mechanical equilibrium, between the thermal comparator and the objects being compared. Some relaxation of these conditions is discussed later.

The foregoing examples present thermal equilibrium as a continuum concept. It can happen that a change in their states can occur when two objects are placed in contact even when both mechanical and thermal equilibrium exists. For instance, molecular diffusion and chemical reactions can take place across the face of contact. Such effects are excluded here.

3.2 Zero'th law of thermodynamics

The observations described, together with others, enable the formulation of what is called the Zero'th Law of Thermodynamics. It can be expressed as follows: if an indicator of fixed state is in thermal equilibrium with two objects each also of fixed state, then the latter objects are in thermal equilibrium with each other.

3.3 The thermometer

Each of the experimental techniques described suggests a design for a thermometer which is basically a thermal comparator. Each design has its disadvantages.

The comparison of colour can readily be performed by using a lamp bulb filament to provide the standard colour and viewing this optically superimposed upon the object under examination. Such an instrument is called the optical pyrometer. Its range is usually limited to colours that are distinguishable by the naked eye.[1] Two parameters which control the lamp's radiance, the temperature of the atmosphere surrounding the instrument and the orientation of the filament affecting its mechanical strain under its own weight,[2] must be controlled or accounted for. This optical pyrometer is particularly useful in that the requirements for a lack of relative motion and presence of mechanical equilibrium mentioned in § 3.1 can be relaxed.

The use of a very thin film of metal to provide an electrical conductivity results in a thermometer which can have a very rapid response with respect to time. It is, however, very sensitive to mechanical strain,[3] and the effect of this, together with the previously mentioned current effects, form limitations to its use.

Using the pressure of a gas as an indicator of thermal equilibrium requires close control of the gas volume; using the measure of the length of a piece of metal introduces the load on the metal as a further variable and requires control of metallic creep to avoid the introduction of time as a further variable; and using the measure of the volume of a quantity of liquid introduces the pressure acting upon the liquid as a variable that affects the thermal equilibrium.

3.4 Temperature

Most thermometers have two principal properties which control the thermal equilibrium. For example: in using a quantity of either air or liquid as the comparator, both its pressure and volume must be fixed; in using a piece of metal both its length and the load acting upon it must be fixed. Call

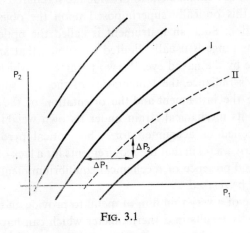

Fig. 3.1

two such properties P_1 and P_2. It is found that thermal equilibrium between an object of fixed state and a thermometer can be obtained for the latter having a range of values of P_1 together with associated values of P_2. An illustration of this relationship is indicated in Fig. 3.1 by the curve denoted I. This line, of a single thermal equilibrium, is called an isotherm. Possession of this result enables the fixed state requirement for the comparator, mentioned in § 3.1, to be relaxed. Equilibrium occurs over a range of values of the state of the comparator.

Were the thermometer to be placed in contact with an object with which it was out of thermal equilibrium, it could be brought to the new level of equilibrium by the property changes $\varDelta P_1$ and $\varDelta P_2$. This change of state is illustrated in Fig. 3.1. Again another line of thermal equilibrium can be determined and such a line is there denoted II. A whole family of these isotherms could be determined as sketched in this figure. The concept of temperature can now be introduced as being the numerical value assigned to each isotherm. Equality of temperature between two objects thus implies thermal equilibrium.

Because temperature is now defined as a function of only properties of the thermometer, it must itself be another property.[†] A thermometer that provides more than one measure for the same temperature has obvious inconveniences. So the choice of material for a thermometer must be made so that the gradient of any isotherm does not change sign along it; also the functional relationship between P_1 and P_2 must be free of discontinuities. Equally, it is important that no two isotherms should intersect; if they did so, the values of P_1 and P_2 at the point of intersection would correspond to two temperatures, and such a thermometer could not distinguish them. The discussion is analogous to that of height in § 1.1; for a thermometer, the temperature is required to form a single valued function of P_1 and P_2.

Practical use requires the choice of a numerical scale of temperature T. A simple choice is to assign to P_2 an arbitrary constant value and to write the linear relation

$$T = a + bP_1. \tag{3.1}$$

Determination of the two coefficients a and b requires two temperatures to be associated with two values of P_1.

An unsatisfactory feature of this method of determining a temperature scale is now apparent: the scale depends upon the physical properties of the material of the thermometer. Transfer of temperature values to another type of thermometer—or

[†] Further application of this statement is deferred until § 5.9.

sometimes even to another thermometer of the same type—usually results in a non-linear scale on the latter. The great advantages of the linear scale in manufacture and in ease of taking readings is thereby lost.[†]

In addition there are obvious advantages in choosing a thermometer design for which variation in P_2 has, at the most, only a very weak effect upon the temperature. That is, in the plot of Fig. 3.1 the isotherms should be as nearly vertical straight lines as possible. The degree of control required upon the value of P_2 is thereby eased.

A thermometer which provides a temperature scale that is independent of its physical characteristics is described later in § 9.3. That it does so is confirmed elsewhere in the present series[‡] where such a scale is defined. This scale is called the absolute temperature.

3.5 Mercury-in-glass thermometer

The mercury-in-glass thermometer is an example of thermometer design. One is illustrated in Fig. 3.2, where the various features are indicated. The property P_1 is identified with the distance along the tube to the position of the meniscus, and P_2 with the mercury vapour pressure in the sealed tube. The physical characteristics which then determine a temperature scale measured by this thermometer would include, amongst others, such features as the thermal expansion characteristics of the mercury, of the glass, and of the mercury vapour, the dimensions of the thermometer, and the surface tension characteristics of the mercury.[4]

[†] The delights of always using a linear scale should not mesmerize one into saying that the second thermometer would give incorrect temperature readings. It would not if it had been correctly calibrated against the first one.

[‡] Montgomery, *Second Law of Thermodynamics*, Pergamon, 1966.

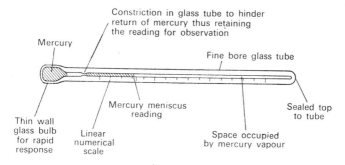

Fig. 3.2

3.6 Continuum limitations of temperature

Sometimes a difficulty arises with the foregoing definition of thermal equilibrium and hence of temperature. An example of this arises with the phenomenon present in the gaseous discharge electric light tube.[5] An optical pyrometer would indicate an extremely high temperature; direct contact with the glass tube would indicate a temperature only slightly above that of the surroundings. The tube contains a mixture of electrons and comparatively much heavier ions. Though they are in contact, and though there is no rate of change of the properties of each of these two constituents with time so that an equilibrium is obtained, their temperatures are greatly different. Such a gas is called a non-isothermal gas. Another example of this kind of phenomenon occurs in the earth's atmosphere at a high altitude where radiation can result in similar effects.[6] In these cases, a continuum thermal equilibrium is not attained. However, these two systems could not continue in the states described if they were completely isolated from their surroundings. For then, in the first case electricity would cease to flow and, in the second, radiation would be cut off.

References

1. KOSTKOWSKI, H. J. and LEE, R. D., Theory and methods of optical pyrometry, *Temperature* (ed. C. M. Herzfeld), Reinhold, 1962, vol. 3, pt. 1, p. 450.
2. *Ibid.*, p. 456.
3. EVANS, J. P. and BURNS, G. W., A study of stability of high temperature platinum resistance thermometers, *loc. cit.*, p. 313.
4. HALL, J. A. and LEAVER, V. M., Some experiments in mercury thermometry, *loc. cit.*, p. 231.
5. VAN ENGEL, A., *Ionized Gases*, 2nd edn., Oxford, 1965, p. 242.
6. SHARP, G. W., The concept of temperature in the upper atmosphere, *Temperature* (ed. C. M. Herzfeld), Reinhold, 1962, vol. 3, pt. 1, p. 819.

CHAPTER 4

UNITS

4.1 Choice of units

Practical reasons require the assignment of numerical values to quantities. The basic standards should be so chosen that they do not vary with time and position; apart from this the choice of size for unit dimension is arbitrary.

4.2 Time

The standard unit of time is called the second. It is based upon the frequency of a certain emission from caesium atoms, and the numerical factor was so chosen, that within the accuracy of measurement when the definition was internationally agreed, the second was made equal to the astronomical second. The length of a day as a unit of time is shown to be unsatisfactory when it is compared with this frequency standard, for between 1956 and 1963 this length varied between $+1\cdot5$ and $-0\cdot5$ msec. Application of the multiplying factors of 60 and 60^2 supplies the definitions of the minute and the hour.†

---

† The apparent contentment of the advocates of the metric system with the subdivision of time by these factors is surprising.

4.3 Length

The metre, as a unit of length, is based upon the wavelength in vacuum of a particular kind of radiation of the krypton-86 atom. The application of a multiplying factor of 10^{-2} forms the definition of the centimetre and the application of a factor of 0·9144[†] provides the definition of the yard. The further factors $\frac{1}{3}$ and $\frac{1}{12}$ can then be applied successively to provide the definition of the foot and the inch.

4.4 Force and mass

The units of time and length having been fixed, the units of force and of mass are directly related to each other through the definition of mass given in eqn. (1.8).

The internationally recognized unit of mass is the International Prototype Kilogram kept at Paris. A smaller unit of mass, the pound, is defined in terms of this kilogram.[†] To five-figure accuracy the conversion factor is 0·45359. A further derived unit of mass is the gram which is defined as 10^{-3} of a kilogram. An independently derived unit of mass is the atomic mass scale[‡] whereby the molar mass of carbon-12 is exactly 12. The molar mass of hydrogen to five figures is then 1·0080. A unit of quantity is the mole, which is defined as that amount whose mass in grams is numerically equal to the atomic mass. There being approximately $6·02 \times 10^{23}$ molecules in one gram mole (a numerical value known as Avogadro's constant), then

---

[†] British Weights and Measures Act of 1963. These definitions of the yard and pound are now universal.

[‡] Erroneously called the atomic weight.

TABLE 4.1

Note: ABBREVIATIONS, WHERE USED, ARE APPENDED TO THE NAME OF THE UNITS

Length	Time	Mass	Force	Work, energy, and heat	Power
metre,[a] m centimetre, cm foot, ft	second, sec second, sec second, sec	kilogram,[b] kg gram, g pound, lb	newton, N dyne poundal	joule,[c] J erg —	watt, W — —
foot, ft	second, sec	slug	pound,[d] lb	—	[e]

[a] The *micron* is defined as 10^{-6} m. The *Ångstrom unit* is defined as 10^{-10} m.
[b] The *tonne* is defined as 10^3 kg.
[c] A unit of heat is the calorie. The different calories were abolished in 1948 in favour of the joule. To the accuracy of measurement, one 15°C g cal equals 4·1855 J. Another unit of heat is the British Thermal Unit. To the accuracy of measurement one 60°F B.t.u. equals 1054·54 J. One Therm is defined as 10^5 B.t.u.
[d] The *ton* is defined as 2240 lb, the *short ton* as 2000 lb.
[e] The *unit horse-power* is defined as 550 ft-lb sec^{-1}.

unit mass on the atomic scale corresponds to

$$\frac{10^{-23}}{6 \cdot 02} = 1 \cdot 66 \times 10^{-24} \text{ g}.$$

The preceding definitions are absolute and quite independent of the accuracy of methods of measurement.[†] They are set out in the first three lines of Table 4.1. The following two definitions, whereby the unit of force is specified and the unit of mass follows from eqn. (1.8), are not independent of other measurements.

The pound is also used as a unit of force defined as the weight force that acts upon a pound mass. Equation (1.11) shows that this definition requires specification of a value of g. At the time of writing there is no general international definition of the value of g to be used, but accepting the authority of Kaye and Laby,[1] the value of $32 \cdot 1740$ ft sec^{-2} will be adopted here. The pound force then gives unit acceleration to a unit mass that is $32 \cdot 1740$ times the pound mass. This unit is called the slug. This system is set out in the fourth line of Table 4.1.

Unfortunately, there is a common and confusing practice of using the pound simultaneously as a unit of mass, written 1bm, and as a unit of force, written 1bf. This is incompatible with the definition of mass given by eqn. (1.8). If the previous definition of the pound force is retained, then in this system

$$32 \cdot 1740 \ F = ma,$$

and the numerical factor has to be carried in all relevant analyses. The various users of this system have adopted the symbols g_o, g_n, and g_c for the factor $32 \cdot 1740$.

[†] Though it cannot be certain that the mass of the International Prototype Kilogram has not been affected by handling over the years.

4.5 Further derived mechanical units

Some mechanical quantities which are based upon the definition of mass, length, and time have had names given to them. They also are tabulated in Table 4.1.

4.6 Temperature

All quantities in mechanics can have their units expressed in terms of only mass, length, and time. Using the appropriate kinetic theory enables the temperature of a gas to be expressed in terms of a mean kinetic energy of the gas molecules. Dimensional analysis would thereby be hampered,[†] and so a unit of temperature is introduced. The internationally recognized unit of temperature is measured by a gas thermometer which is described in § 9.3. The corresponding scale is defined by putting $a = 0$ in eqn. (3.1) and making the temperature T at the triple point of water[‡] have the value of 273·16. The units are called degrees Kelvin, written °K.

From this fundamental scale other scales can be derived. Temperature t on the Celsius scale is related to temperature T on the Kelvin scale by[§]

$$t = T - 273 \cdot 15,$$

where the temperature of the ice point of water is 273·15° K.

[†] See ref. 2 and also P. Bradshaw, *Experimental Fluid Mechanics*, Pergamon, 1964.

[‡] The triple point of water is that state where H_2O coexists in liquid, solid, and vapour form.

[§] The term Centigrade, which was in use in Great Britain and the United States of America, was dropped in 1948 on international agreement being reached upon this definition of the Celsius scale. In the metric system, unit centigrade is 10^{-4} of 90° of angle.

Two other temperature scales which are occasionally used are the Rankine and Fahrenheit scales. No internationally accepted definition of these scales exist. Common practice[3] is to use the following definitions:

Temperature in degrees Rankine = $\frac{9}{5} \times$ (temperature in degrees Kelvin).

Temperature in degrees Fahrenheit = $32 + \frac{9}{5} \times$ (temperature in degrees Celsius).

4.7 Current

Electrical quantities require a further unit to describe them. The internationally accepted unit is the ampere, defined as that current in two straight and parallel conductors of infinite length and zero diameter, which are one metre apart and which in vacuum exert a force of $2 \cdot 10^{-7}$ newton per metre length.[†]

4.8 Light

The final of the six basic units, the candela, is not required in the present studies.

References

1. KAYE, G. W. C. and LABY, T. H., *Tables of Physical and Chemical Constants and some Mathematical Functions*, 12th edn., Longmans, London, November 1960, p. 9.
2. SEDOV, A. I., *Similarity and Dimensional Methods in Mechanics*, Infosearch, London, 1959, p. 41.
3. *Steam Tables 1964*, National Engineering Laboratory, HMSO, Edinburgh, 1964 (prep. R. W. Bain), p. 144.

† There is a possibility that in the future the ampere may be defined in terms of the gyromagnetic ratio of the proton.

CHAPTER 5

THE SYSTEM OF FINITE SIZE

5.1 The system

The characteristics of infinitesimal elements have been discussed in Chapter 2. An introduction to those of finite volumes of material has been given in Chapter 3 by a description of examples and with the introduction of terms such as the property. Discussion of the finite volume of material now continues with a more rigorous definition of these terms.

A collection of material is called a system. During a process the material of which the system is composed may be redistributed within the boundary of the system; the system boundary may change its shape; the volume enclosed by the system boundary may change its value; but it is a requirement of the definition that no material is transported across the system boundary. It is not enough that the mass of the system should remain constant, for two separate but equal amounts of mass of material could enter and leave a volume during a process, and the volume would not then have contained a system. This definition excludes molecular and atomic motions across a boundary by either the process of diffusion under a concentration gradient, or by the mobility of electrically charged particles under an electric field.

The state of a system is determined by the observation of its characteristics, here called its properties. The ability to distinguish a change in the state of a system is limited by the ability to make experimental observations of its property changes. If, for example, only mass could be observed, it would not be possible to distinguish between a gram of ice and a gram of water. A sufficient number of properties must be evaluated to distinguish fully the change in the state of a system that results from a process.

The initial discussion of this chapter on the properties of finite systems is limited to systems that are in equilibrium, both mechanical and thermal.

5.2 The stress in a solid

Within a solid, three mutually perpendicular planes passing through a point can be so orientated that at this point there are no tangential stresses lying within any of these principal planes.[1] In general, on any other planes there will be both normal and tangential stresses acting at the point. Thus to define the state of a solid system at least it is necessary to specify the direction of the principal planes and the principal normal stresses acting upon them together with the distribution of these properties throughout the solid. Only in very simple cases, such as when a cylindrical bar is in equilibrium under a uniformly distributed tensile stress, can these properties be reduced to one in number which is single valued throughout the system.

5.3 Tangential stress in liquids

When a liquid such as water flows steadily past a smooth solid surface, experiment indicates four features of such a flow. First, if measurements of the velocity are made sufficiently closely to the wall, a distribution is obtained as shown in

Fig. 5.1,[2] indicating that, if extrapolated, the velocity tends to zero at the wall: second a shear stress τ_0 is found to act upon the wall: third, the direction in which the velocity vector q tends to at the wall, is the direction of the stress vector τ_0: and fourth, if y is the distance from the wall, then the properties

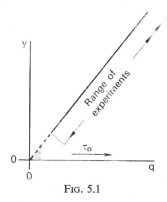

FIG. 5.1

of the liquid at the wall being constant, the velocity gradient at the wall $(dq/dy)_{y=0}$ and τ_0 are related by

$$\tau_0/(dq/dy)_{y=0} = \text{a positive constant.}$$

This constant is called the coefficient of viscosity μ and is a property of the liquid. So that

$$\tau \equiv \mu(dq/dy)_{y=0}. \qquad (5.1)$$

Liquids which obey this relation are called Newtonian fluids. There is an upper limit to the shear stress which a liquid can sustain. At sufficiently high rates of shear, that is of (dq/dy), a liquid can rupture.[3]

As a corollary of eqn. (5.1) it is seen that within a motionless Newtonian fluid there are no tangential stresses at a solid boundary; in fact, as will be discussed later, there are no tangential stresses at any point within it. Conversely, application of a tangential stress to a fluid, whether Newtonian or not, results in motion. This is a mark that distinguishes a fluid from a solid.

The value of μ is not necessarily constant throughout a flow; at any point it depends upon the local state of the liquid.

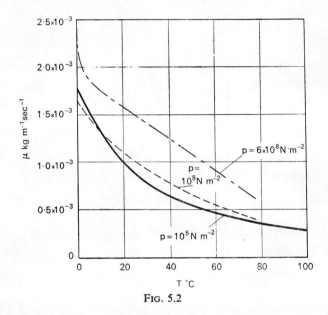

Fig. 5.2

Measurements of the viscosity of liquids show that it is a function of only the temperature except at extremely large pressures. Measured values for water taken from refs. 4 and 5 are illustrated in Fig. 5.2.

Many liquids do not obey the linear relation between the stress and the velocity gradient given by eqn. (5.1).[6] The effect of shearing the flow of a liquid composed of long-chain molecules is to give a tendency to the molecules to align themselves with the direction of the shear. This means that the viscosity varies with direction and also that the shear is not linearly proportional to the velocity gradient. Very often such liquids satisfy the relation

$$\tau_0 \propto \left(dq/dy\right)^n_{y=0}, \qquad (5.2)$$

where the index n and, as before, the constant of proportionality are properties of the liquid. These liquids are described as non-Newtonian. Liquids which normally act as a Newtonian fluid can become non-Newtonian in their behaviour at very high rates of shear.[7]

Other materials only behave as fluids under limiting conditions; in some cases they do so above a limiting value of the strain; in others they do so below a limiting value of the rate of strain. More complex effects occur when a liquid is composed of dipole molecules and electrical force fields act.[8]

5.4 Tangential stress in gases

Liquids and gases are classed under the general title of fluids; usually the behaviour of the latter in motion is that of a Newtonian fluid obeying eqn. (5.1). However, when a gas has a very low density the molecular mean free path can become comparable with, or greater than, the length $q_s/(dq/dy)_{y=0}$, where q_s is a typical general velocity of the flow: the corresponding Knudsen number is unity. Then the gas velocity noticeably does not tend to zero at the wall; there is a slipping effect which, for the experiments illustrated in Fig. 2.1, is significant for Knudsen numbers above 0·02 in the correspondingly called slip flow regime.[9]

5.5 Variation of pressure in a stationary fluid

Figure 5.3 is a sketch of a horizontal cylinder drawn in a fluid that is at rest under gravity.† Thus from the discussion of §§. 5.4 and 2.5 the tangential stress is everywhere zero and the reverse of the normal stress can be identified with the pressure. The cross-sectional shape and size of the cylinder is constant

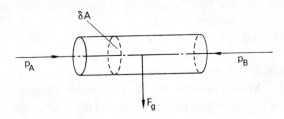

Fig. 5.3

along its length, the corresponding area being δA. On its end planes, which are vertical ones, the pressures p_A and p_B are acting. If δA tends to zero, at this limit the variations of p across it will be of no consequence, and so the resultant horizontal force acting upon the cylinder will be

$$p_A \, \delta A - p_B \, \delta A.$$

As there is no acceleration present this force must be zero and so

$$p_A = p_B,$$

a result that is independent both of the length of the cylinder and of its orientation in a horizontal plane. Thus the pressure

---

† The *Oxford Dictionary* definition of horizontal is not precise in the present context. Here "vertical" is defined as tangential to the local direction of the gravitational force, and "horizontal" is defined as being perpendicular to vertical.

THE SYSTEM OF FINITE SIZE

is constant at all points on a horizontal plane in a stationary fluid under gravity.

A sketch of a vertical cylinder of small length δz, where z is the altitude above some arbitrarily chosen level, is shown in Fig. 5·4. On the lower end a pressure p is acting, on the upper acts a pressure $p + \delta p$.

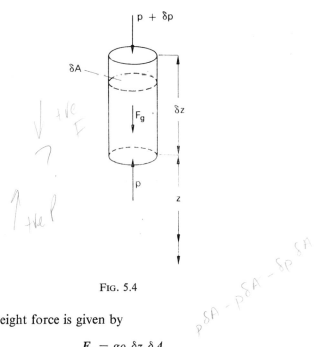

Fig. 5.4

The weight force is given by

$$F_g = g\varrho\, \delta z\, \delta A,$$

whilst the vertical pressure force is

$$p\, \delta A - (p + \delta p)\, \delta A = -\delta p\, \delta A.$$

Thus, equating these two forces,

$$g\varrho\, \delta z = -\delta p.$$

The change in pressure δp occurs in the z-direction so that as $\delta z \to 0$ it can be expressed by

$$\delta p = \frac{dp}{dz}\,\delta z.$$

Eliminating δp between these last two relations gives

$$\frac{dp}{dz} = -\varrho g. \tag{5.3}$$

Thus the pressure in a stationary fluid system decreases with altitude at a rate that is proportional to the fluid density.

For a small change in height δz, the percentage change in pressure is

$$\frac{\delta p}{p} = -\frac{\varrho}{p} g\,\delta z.$$

Taking water as a typical liquid, and atmospheric pressure as a typical pressure, then for this numerical example,

$$\varrho = 10^3 \text{ kg m}^{-3},$$
$$p = 10^5 \text{ N m}^{-2},$$
$$g = 9\cdot 8 \text{ m sec}^{-2},$$

So that approximately,

$$\frac{\delta p}{p} = -0\cdot 1\,\delta z$$

where δz is in metres.

Or, again for air at normal atmospheric conditions,

$$\varrho = 2\cdot 4 \times 10^{-3} \text{ slug ft}^{-3},$$
$$p = 2\cdot 1 \times 10^3 \text{ lb ft}^{-2},$$
$$g = 32 \text{ ft sec}^{-2},$$

then $\quad \dfrac{\delta p}{p} = -0\cdot 4 \times 10^{-4}\,\delta z,$

where δz is in feet.

These examples illustrate the fact that in most liquid systems the fractional variation of pressure through the system due to the presence of gravity is significant whilst in a gas it is very small. In making observations of a stationary gaseous system a single value of the pressure can be regarded as a property of that system. The pressure used to define the state of a liquid must be more carefully specified; it is usual, where applicable, to adopt the value of the pressure at a free surface, that is in contact with the surrounding atmosphere.

5.6 Continuity of stress

There can be no discontinuity in the distribution of stress within a continuum system that is composed of material in only one of the solid, liquid and gaseous states. For a thin slice of material at such a discontinuity would experience a finite force and hence an infinite acceleration as the slice thickness tended to zero.

The stress can also be taken as continuous at boundaries between these types of system except at the liquid–gas boundary where a significant surface tension exists. The balance of stress in this latter case is discussed in § 10.2.

At any point in a system the body force on a small element of that system is proportional to the mass of the element. Hence discontinuities in the distribution of this force per unit mass are permissible. For example, a discontinuity can occur at the boundary between a system of magnetic material and non-magnetic surroundings.

5.7 Temperature

If a system is split up into small infinitesimal ones, each being in thermal equilibrium with its neighbour, then there is thermal equilibrium throughout the finite size system. It then follows that the temperature has a constant value throughout and it can, under this equilibrium condition, be regarded as a single valued property of the whole system.

5.8 State and property changes

The discussion of the preceeding sections indicates that a quantity of gas which is stationary and in thermal equilibrium within itself has single values of the pressure and temperature. This condition of mechanical and thermal equilibrium is called

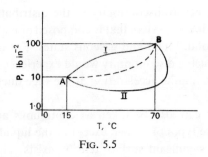

Fig. 5.5

an equilibrium state. As an example these pressure and temperature values might be 10 lb in^{-2} and 15° C, associated with, say, a state A. On a graph of pressure against temperature, such as is sketched in Fig. 5.5, this state is indicated by the point A. A process might take place so that the system changes to a new equilibrium state indicated by B where the two properties, for example, now have the values 100 lb in^{-2} and 70° C. The change of state is recognized by the change in the properties.

At any stage during the process the pressure and temperature would not be expected to be uniform throughout the system; an arbitrarily chosen pair of elements would be likely to have different associated pressures and temperatures.† The values of p and T for one element as they vary during the process might plot as the curve marked I in Fig. 5.5, and those for the other as curve II. Such curves, representing the continuous effect of a process upon the properties of an element, are called paths.

A path can be chosen from amongst the infinite number for all the infinitesimal elements taking part in a particular process, or indeed the whole nature of the process from A to B can have an infinity of variations. If the end states are fixed at A and B, the total change in pressure and temperature is independent of the choice of path.

The discussion is analogous to that of height in § 1.1. If discontinuities are excluded, then pressure and temperature can be regarded as perfect differentials so that

$$p_B - p_A = \int_A^B dp$$

and

$$T_B - T_A = \int_A^B dT.$$

A unique state is known to be so by observing that each of the corresponding properties has a certain value. Thus the foregoing arguments apply to all properties: excluding discontinuities, all are perfect differentials.

Also, a quantity whose change during a process depends only upon the end states and is independent of the path can be regarded as a property provided an origin from which its values are measured is defined. For then it will have a unique numerical value for a particular state.

Variations of the properties throughout a system undergoing a process are sometimes negligible compared with the changes

---

† In so far as pressure and temperature can be defined for an element in motion (see § 11.9 later).

resulting from the process so that the change of state can then be indicated, on a plot of properties, by a single line such as that shown dotted in Fig. 5.5. Always such an approximation must be justified for the process to which it is applied.

5.9 Classification and derivation of properties

Properties have different characteristics which enable them to be classified into three divisions as now tabulated.

(a) When a system is in thermal equilibrium with its surroundings and within itself, the temperature is uniform. When a system comprised of only fluid is in mechanical equilibrium, the pressure is similarly uniform and this also occurs in the simple stressing of a solid described in § 5.2. When an electrically conducting solid system is in electrical equilibrium, then the voltage is uniform. In these cases the pressure, the stress, the temperature, and the voltage are homogeneous properties.

When a property is not homogeneous, it can be classified as heterogeneous in two divisions.

(b) The mass of a system is formed by the sum of the masses of its parts, so also is the volume determined.[†] These properties are additive ones. Care must be taken in applying these results. For example a system can be formed by the mixing of two substances resulting in spontaneous changes occurring in the mixture, so that, for instance, the final volume, is not the sum of those of the two constituents.[‡]

[†] The second of these results follows from the way in which volume is measured, the first is demonstrated later in § 12.1.

[‡] See the further discussion in S. R. Montgomery, *Second Law of Thermodynamics*, Pergamon, 1966 (in the same series), p. 100.

THE SYSTEM OF FINITE SIZE

(c) The density is a non-additive property. When, for example, the masses and volumes of the two parts of a system are m_1, m_2, V_1, and V_2 respectively then the mean density ϱ is,

$$\varrho = \frac{m_1+m_2}{V_1+V_2}$$

which is not equal to $\dfrac{m_1}{V_1}+\dfrac{m_2}{V_2}$.

A single-valued function of properties has a unique value for any particular state of a system. The arguments of § 5.8 show that this type of function is also a property.

The addition of two additive properties forms a new property that is also an additive property. For example, if α and β are dimensional constants,

$$\alpha m + \beta V = \alpha m_1 + \alpha m_2 + \beta V_1 + \beta V_2$$
$$= (\alpha m_1 + \beta V_1) + (\alpha m_2 + \beta V_2),$$

and so the property $(\alpha m + \beta V)$ is formed by the addition of its parts.

The product of a single valued function of a homogeneous property and an additive property is another additive property as illustrated by

$$V_1 f(p) = (V_1+V_2)f(p)$$
$$= V_1 f(p) + V_2 f(p).$$

An additive property has another important characteristic. Consider the additive property X, say, of a homogeneous system of mass m. If the system is split into two halves, then X will be formed by the sum of the corresponding property of each half which by symmetry must be $X/2$. So also the mass of each half will be $m/2$. The ratio of the property of each portion to the mass of that portion is

$$\frac{X/2}{m/2} = \frac{X}{m} = \text{constant}.$$

Successive division of the portions of the system by two, will form any proportion of the whole system, and thus for any proportion the value of the property divided by the mass is a constant, or generally

$$X \propto m.$$

An additive property of a homogeneous system is thus proportional to the system mass and so there is convenience in defining a value of the property per mass unit, x say, by

$$x \equiv \frac{X}{m}, \qquad (5.4)$$

and x is then a homogeneous property, but it is not an additive property as X is.

The properties considered here have four different associations as follows:

(a) The geometric properties are the space coordinates x, y, and z and the time t.

(b) The mechanical properties include the velocity q and the acceleration $\dot{q}$.

(c) The electrical properties include the ion concentration, the charge, and the voltage.

(d) The thermostatic properties include the stresses σ, τ, and p, the density ϱ, the temperature T, and the viscosity μ.

5.10 Work

The definition of work given in § 1.7 implies a sign convention. If the object is a system and a force acting on it moves in its own direction then a positive quantity of work is done to the system; equally, if the force is a surface one, a negative amount of work is then done by the system upon its surroundings.†

When a system consists of an engine, interest is centred upon the usefulness of the work done upon the surroundings. Such usefulness could be indicated by the ability of the engine to wind up an elastic spring, to raise a weight force through

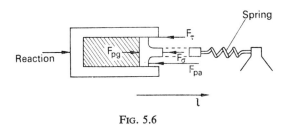

Fig. 5.6

a gravitational field, and by the ability to move an electrical conductor carrying a current through an electrical force field.

The total force exerted by a system upon its surroundings is divided into three components, that due to the normal boundary stress σ, that due to the tangential boundary stress τ, and that due to the body force per unit mass β. The ability of a system to do useful work upon its surroundings by the media of these three force components is now discussed.

Figure 5.6 shows a quantity of gas, at a pressure that is higher than atmospheric pressure, confined within a cylinder by a piston. The piston, on being released, would move horizontally and a process, in which a spring could be compressed, would ensue. The forces, indicated in this figure, that act upon the piston and rod are that due to the gas pressure on the piston F_{pg}, that due to atmospheric pressure F_{pa}, that due to friction between the piston and cylinder F_τ, and, finally, that due to the compressive stress in the rod F_σ.

† A conceptual difficulty arises in the case of surfaces that appear to slip over each other. This is discussed later in § 8.1.

In a process from state A to state B the work done by the gas to the piston W_g is given by

$$W_g = \int_A^B F_{pg}\, dl. \qquad (5.5)$$

The work done to the rod by the spring will be $-\int_A^B F_\sigma\, dl$ so that done to the spring W_r will be

$$W_r = \int_A^B F_\sigma\, dl. \qquad (5.6)$$

The total horizontal force F on the piston and rod at any instant during the process will be given by

$$F = F_{pg} - F_{pa} - F_\tau - F_\sigma,$$

and so the total work done on the piston and rod $\left|_A^B\right. W$ will be given by[†]

$$\left|_A^B\right. W = \int_A^B (F_{pg} - F_{pa} - F_\tau - F_\sigma)\, dl.$$

If at the beginning and end of the process the piston and rod assembly is stationary, then, by eqn. (1.18),

$$\left|_A^B\right. W = 0,$$

and, on substituting from eqns. (5.5) and (5.6),

$$W_g - W_r = \int_A^B F_{pa}\, dl + \int_A^B F_\tau\, dl. \qquad (5.7)$$

On the right-hand side of this equation both F_{pa} and F_τ as drawn in Fig. 5.6, are positive, and so

$$W_g > W_r.$$

[†] § 8.1 explains why this is only true if the film of lubricant is not included with the piston.

If the work done by a system is to be measured by the former criterion of usefulness, then the work done to a system that comprises the spring by all its surroundings will be W_r. It might be thought that the useful work done by a system consisting of just the gas is still W_r, and this, from above, is less than W_g. However, a system is thought of as having an absolute ability to do useful work that is not to be penalized by failings in perfection of its surroundings. The work done by the gas is W_g, this being the maximum useful work that could be done by the gas system if the atmospheric pressure and the piston friction were reduced to zero.

In this discussion a reaction force to retain the cylinder in position is permitted. There is no contribution to the work done on the surroundings because the cylinder support is assumed to be motionless. This example shows the importance of taking care in defining a system before the work done to or by it is evaluated.

Similarly, a shear force at a moving boundary of a system can do useful work. An apparatus whereby this might be done is illustrated in Fig. 5.7. Liquid flowing under gravity between two tanks passes and exerts a shear force upon the surface of a rotating cylinder. The torque so generated could do useful work by the winding of a string on to a drum and the raising of a mass against a gravitational weight force. Again, imperfections in the surroundings, such as friction at the liquid seals, which hinder the conversion of the torque exerted by the liquid into useful work, are not debited against a system comprised of the liquid.

Again, a body force such as exists between two masses or between the poles of two magnets can do useful work. A suitable arrangement for doing this is sketched in Fig. 5.8. If the system comprises a permanent magnet attached to a spring which is compressed against a stop, then on release of the stop the spring will expand and the magnet will move to the right

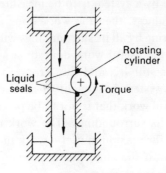

Fig. 5.7

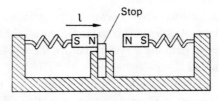

Fig. 5.8

as shown, thereby changing the force it exerts upon the second magnet. Depending upon the alignment of the magnetic poles, this would then compress or extend the second spring outside the system, thus doing useful work.

5.11 Evaluation of work

The work done to a system during a process, given by eqn. (1.16), is the sum of its three parts; the work done by the normal boundary stresses,

$$\left. W_\sigma \right|_A^B = \int_A^B \dot{W}_\sigma \, dt;$$

that done by the tangential boundary stresses,

$$\left| ^B_A W_\tau = \int_A^B \dot{W}_\tau \, dt; \right.$$

and that done by the body forces,

$$\left| ^B_A W_\beta = \int_A^B \dot{W}_\beta \, dt. \right.$$

Summing these three contributions to the work and noting eqn. (1.16), gives

$$\int_A^B \dot{W} \, dt = \int_A^B (\dot{W}_\sigma + \dot{W}_\tau + \dot{W}_\beta) \, dt,$$

and so $\dot{W}$ can be identified with the integrand on the right-hand side giving

$$\dot{W} = \dot{W}_\sigma + \dot{W}_\tau + \dot{W}_\beta. \qquad (5.8)$$

Figure 5.9 is a sketch of a portion of the boundary of a system of small area da, and upon which the normal and tangen-

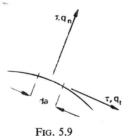

FIG. 5.9

tial stresses σ and τ are acting. The components of the boundary velocity in the direction of these stresses are written q_n and q_t respectively.[†]

The rate at which work is being done to the system by the

[†] It should be observed that there may also be a velocity component that is perpendicular to both σ and τ.

normal stress acting over the area da is then

$$\sigma \, da \, q_n. \tag{5.9}$$

The total rate of doing work to the system is the summation of all such elemental contributions taken over its surface. This is written

$$\dot{W}_\sigma = \int \sigma q_n \, da. \tag{5.10}$$

Similarly,

$$\dot{W}_\tau = \int \tau q_t \, da.$$

If the body force per unit mass is written β, then for every element of mass dm within the system the rate of doing work to the system is

$$\beta \, dm \, q_\beta,$$

where q_β is the component of the velocity of the element in the direction of the body force. These elemental rates of doing work have then to be summed over the whole volume of the system, and so

$$\dot{W}_\beta = \int \beta q_\beta \, dm \tag{5.11}$$

Thus by noting eqns. 1.16 and 5.8, the evaluation of the work done during a process requires a double integration and so

$$W = \iint (\sigma q_n + \tau q_t) \, da \, dt \\ + \iint \beta q_\beta \, dm \, dt. \tag{5.12}$$

Equation (5.9) can be expressed in another way. In time δt, δa moves outwards a distance q_n. The corresponding increase in volume at this part of the system boundary ∞V is thus $q_n \, \delta a$.[†] Thus the work done to the system in this time can be expressed by

$$\sigma \infty V. \tag{5.13}$$

[†] Requirements of §§ 8.3 and 13.2 force the use of another symbol to denote a differential.

5.12 Heat

A lack of mechanical equilibrium results in work being done by the unbalanced force. A lack of thermal equilibrium results in heat being applied by the unbalanced temperature. Heat is the interaction between a system and its surroundings due to the presence of a temperature difference between the two at the instant before they are brought into contact.

This is illustrated by an experiment whereby 1 kg of aluminium having a temperature of 60° C is suspended within 1 kg of water at a temperature of 30° C. A process would then ensue. Eventually thermal equilibrium between the water and the aluminium would be achieved and, provided the heat interaction is only between the water and the aluminium, their final common temperature would be about 35·2° C. If this process is to be one of heat alone, then a refinement to this description is required to ensure that no work occurs at the boundary of the water. Conceivably this could be done by suspending the aluminium so that contact is just maintained with the free surface of the water which, being exposed to a perfect vacuum, is at zero pressure. Also the centre of gravity of the water must be held still.

This experiment could be repeated m times, the temperature of the water would rise 5·2° C each time; the total amount of heat applied would be m times that in the initial experiment. Or, a single experiment using m kg of water and m kg of aluminium could be used; again the heat applied would be m times that applied in the initial experiment. Thus the mass of water m, whose temperature rises from 30° C to 35·2° C, could be used as a measure of the heat. Such a unit of heat is called a calorie. One example of a calorie is the 15° C g cal which is definable as the heat which raises the temperature of 1 g of water from 14·5° C to 15·5° C at a standard atmospheric pressure. Another is the amount of heat which similarly raises 1 lb mass of water from a temperature of 59·5° F to one of 60·5° F; this latter unit is called the 60° F lb cal, or British thermal unit. And another is the amount of heat which raises 1 lb mass of water from a temperature of 14·5° C to 15·5° C; this unit is called the Centigrade heat unit.

Heat has similarities to work whose characteristics were discussed in § 1.7. It is something that happens during a process; it is transient in nature: after a process is completed, heat no longer exists. It is not something that is contained in a system, and so an element of heat applied during a portion of a process is not written dQ, for this would tempt one to write

$$Q_B - Q_A = \int_A^B dQ,$$

and the left-hand side of this relation would be meaningless.

It is important that there should not be confusion between the terms "temperature" and "heat" as used here. The term "cold" in common parlance can cause confusion as witness the following conversation:[10]

> "...and sometimes in the winter we take as many as five nights together—for warmth, you know."
> "Are five nights warmer than one night, then?", Alice ventured to ask.
> "Five times as warm, of course."
> "But they should be five times as *cold* by the same rule...."
> "Just so!", cried the Red Queen.

5.13 Algebraic characteristics of heat

An experiment is illustrated in Fig. 5.10. In it an object at an initial temperature T_0 is immersed in water of mass m g and at a lower initial temperature of 14·5° C. Heat Q is applied by the object to the water during a time chosen to be less than

THE SYSTEM OF FINITE SIZE

that for the two to reach thermal equilibrium but just sufficient for the water to achieve an equilibrium temperature of 15·5° C after contact has been broken. A series of such experiments in which a range of values of the water mass is used gives a corresponding range of ΔT, the temperature drop of the object. Thus
$$\Delta T = f(m).$$

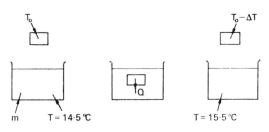

Fig. 5.10

But m has been adopted as the measure of heat so that, with a factor of proportionality k,
$$Q = km. \tag{5.14}$$
Noting that k is a constant, these two relations give
$$\Delta T = f(Q), \tag{5.15}$$
so that, for this experiment, the fall in temperature ΔT is a secondary measure of the heat applied by the object.

Another possible experiment is illustrated in Fig. 5.11. In its first part, two objects, 1 and 2, are each placed in a quantity of water of mass m_1 and m_2 respectively, so chosen that the water has the requisite temperature rise from 14·5° C to 15·5° C, where thermal equilibrium exists. As a result of the heat applications, Q_1 and Q_2, the objects undergo a temperature change of ΔT_1 and ΔT_2.

If instead the two objects, starting from their same initial states, are placed, either together or successively, into a single quantity of water of mass m and if the mass m is so chosen that the requisite rise in temperature from $14\cdot5°$ C to equilibrium at $15\cdot5°$ C is again obtained, then measurement would show that

$$m = m_1 + m_2.$$

As the end temperatures of the process are unchanged, then

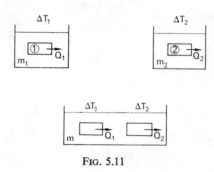

Fig. 5.11

ΔT_1 and ΔT_2 are as before. Hence from eqn. (5.15) so also are Q_1 and Q_2. In the second part of the experiment, from eqn. (5.14),

$$\begin{aligned} Q &= km \\ &= k(m_1+m_2) \\ &= km_1 + km_2 \\ &= Q_1 + Q_2. \end{aligned}$$

Thus heat is simply additive either over two portions of surfaces or successively in time. This is a consequence of the arbitrary choice of the means of measurement of heat.

The experiment just described could be amended, as illustrated in Fig. 5.12, by first placing the two objects in contact and allowing them to reach equilibrium with each other. During this process object 1 would apply heat of amount Q_A to object 2 and the latter would apply Q_B to the former. On being then separated and placed into water as in the second part of the experiment illustrated in Fig. 5.11, the value of m would be found to be as before. The values of ΔT_1 and ΔT_2 being as before then so also are Q_1 and Q_2 over the whole process. Heat being simply additive and the heat amounts in the first part of this experiment being Q_A and Q_B for the two objects, then, the heat amounts in the second part are Q_1-Q_A and Q_2-Q_B. The total amount of heat in the second part is thus

$$Q_1-Q_A+Q_2-Q_B.$$

But m is the measure of this heat which is thus still Q_1+Q_2.

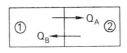

Fig. 5.12

So that equating these two quantities,

$$Q_1-Q_A+Q_2-Q_B = Q_1+Q_2,$$
and thus
$$Q_A = -Q_B, \tag{5.16}$$

showing that heat, like work by surface forces, has an equal and coincident reaction. When heat is by radiation, it is shown in § 7.3 that care must be taken in interpreting this result.

A further deduction can be made from this experiment. If, of their initial temperatures, object 1 has the higher of the two, then Q_A will be associated with a temperature drop and so can be measured as already described. Then Q_B, which can then be computed from eqn. (5.15), is associated with a temperature rise of object 2: for example, this rise might be from, say, 100–105° C. A heat Q'_B, associated with a drop in temperature of this object from 105° C to 100° C, could be measured also in the usual way and it would be found that

$$Q'_B = -Q_B.$$

Thus, in this experiment, a reversal of a temperature drop ΔT is a measure of a precisely equal reversal of heat.

5.14 Evaluation of heat

Heat is a phenomenon that takes time and, as just demonstrated, it can be evaluated by a summation over time. The rate at which heat is being applied being written $\dot{Q}$, then

$$Q = \int \dot{Q}\, dt. \tag{5.17}$$

It also being summable over the system boundary, then a heat rate intensity η can be defined. Referring to Fig. 5.13,

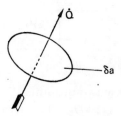

Fig. 5.13

if $\dot{Q}$ is the heat rate acting upon an area δa, then this intensity can be defined as the ratio $\dot{Q}/\delta a$. Heat being a molecular interaction, the limit to which δa can tend for a gas is dependent upon the molecular mean free path L.[†]

Bearing this in mind, the definition is

$$\eta \equiv \dot{Q}/\delta a. \tag{5.18}$$

The heat can then be summed by the relation

$$Q = \iint \eta\, da\, dt. \tag{5.19}$$

In the previous section it was demonstrated that Q has an equal and coincident reaction. As a process can be terminated at any time, then it follows from eqn. (5.17) that so also must $\dot{Q}$ have an equal and coincident reaction.

[†] See also the note on ref. 3 in § 2.1.

5.15 The equilibrium state

By the meaning of an equilibrium state it follows that every part of a system is motionless in the chosen system of axes, and so

$$\dot{W} = 0.$$

Also the temperature is uniform throughout the system and in balance with that of the surroundings, so that

$$\dot{Q} = 0.$$

It thus follows from eqns. (1.15) and (5.17) that whilst an equilibrium state exists,

$$W = 0 \quad \text{and} \quad Q = 0.$$

References

1. TIMOSHENKO, S., *Theory of Elasticity*, McGraw-Hill, New York, 1934, § 53.
2. ELRICK, R. M. II, and EMRICH, R. J., Tracer study of pipe flow behaviour to within two microns of the wall, *Phys. Fluids* **9** (1), 28 (Jan. 1966).
3. HUTTON, J. F., The fracture of liquids in shear: the effects of size and shape, *Proc. Roy. Soc.* A, **287**, 222 (1965).
4. KAYE, G. W. C. and LABY, T. H., *Tables of Physical and Chemical Constants and some Mathematical unctions*, 12th edn., Longmans, London, 1960, p. 36.
5. *Ibid.*, p. 37.
6. WILKINSON, W. L., *Non-Newtonian Fluids*, Pergamon, Oxford, 1960, pp. 4 and 31.
7. MARRIS, A. W., *The Hypo-fluent Liquid*, Developments in Theoretical and Applied Mechanics, vol. 2 (Ed. W. A. Shaw), Pergamon, Oxford, 1965, p. 297.
8. MIESOWICZ, M., The three coefficients of viscosity of anisotropic liquids, *Nature* **158** (4001), 27 (6 July 1946).
9. PATTERSON, G. N., *Molecular Flow of Gases*, Wiley, New York, 1956, p. 183.
10. CARROLL, L., *Through the Looking-glass and what Alice found there*, Macmillan, London, 1965, ch. 9, p. 204.

CHAPTER 6

THE FIRST LAW OF THERMODYNAMICS

6.1 Cyclic processes

The dotted line of Fig. 6.1 marks the perimeter of a system which includes a chamber containing gas in which a paddle wheel can be rotated. Across the boundary of the system work can be done through the rotation of the paddle shaft by a motor, and heat can be applied to a heat sink whose temperature is initially different from that of the system. An initial state of this system could be a motionless one of uniform temperature so that, both the paddle wheel and gas being stationary, the latter would be at a uniform pressure. This state is called state A. Work could be done to the system by driving the paddle wheel for a time and then allowing all motion to subside. The resulting new state is denoted state B. If, as sketched in Fig. 6.1, the paddle blades are flat and of negligible thickness, then the resultant velocity at any point on them has a direction that is perpendicular to their surface and so the work is done to the gas by only the normal stresses.

At the second state the temperature of the system would be higher than at state A. Allowing the system to interact with a heat sink of lower temperature, heat could be applied by the system of such an amount that the system could return again to a stationary state of uniform temperature and pressure identical to the initial one: only time would have changed.

THE FIRST LAW OF THERMODYNAMICS

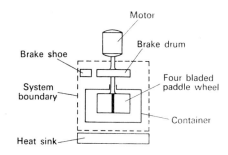

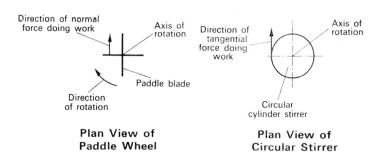

Fig. 6.1

Such a complete process from state *A* to state *B* and back to state *A* is called a cyclic process.

This cyclic process could be modified by replacing the paddle wheel by a stirrer of cylindrical shape and, for instance, placing liquid instead of gas in the container. On rotating the cylinder, work would now be done by only viscous tangential stresses. Again, a cyclic process could be performed by doing work to the system and applying heat to the surroundings.

Another modification to the process would be the replacement of the paddle wheel by a spring which is coiled by the motor. On the latter being disconnected from the system, and the spring being allowed to unwind under the restraint of a brake which also is within the system, then again the temperature of the latter would rise and again the cyclic process could be completed by application of heat to the surroundings.

Or, as illustrated in Fig. 6.2, a system could consist of the gas compressed within a cylinder by a piston held in position by a stop. A cyclic process could be achieved by removing the first stop and allowing the piston to travel freely until it hits the second one, and then doing work in compressing the gas

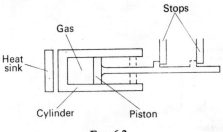

Fig. 6.2

again until the first stop can be replaced, heat finally being applied to a heat sink. When the gas is again stationary and its temperature is uniform and equal to the original value, the cyclic process has been completed.

6.2 First law of thermodynamics

Many such cyclic processes which involve work to and heat by a system could be performed; the measurement of these quantities having been specified respectively in §§. 5.11 and 5.12; their values can be compared giving the following result. The ratio of the work done to $\oint W_{to}$ to the heat applied by $\oint Q_{by}$ is a positive constant for all processes; or, in symbolic form,

$$(\oint W_{to})/(\oint Q_{by}) = J, \qquad (6.1)$$

where J is called the mechanical equivalent of heat.

THE FIRST LAW OF THERMODYNAMICS 69

This is a formulation of what is called the First Law of Thermodynamics.

6.3 The measure of J

The numerical value of the constant J depends solely upon the units used in measuring work and heat; it is a units conversion factor. Carefully performed experiments have resulted in the following values for J;[†]

$$1\ 15°\ \text{C g cal} = 4\cdot1855\ J.$$
$$1\ 60°\ \text{F lb cal} = 778\cdot0\ \text{ft-lb}.$$

Other values are given in Table 4.1 (p. 35).

It is convenient to measure both work and heat in the same units. This gives to J a value of unity. Then eqn. (6.1) becomes

$$\oint W_{\text{to}} = \oint Q_{\text{by}}. \tag{6.2}$$

6.4 Internal energy

Figure 6.3 is a diagramatic representation, on a graph of two properties P_1 and P_2, of two cyclic processes. The first is by path I from state A to state B with a return to state A by path III; the second is by another outward path II from state A to state B again with a return by path III. The single arrows are not to be construed to mean that P_1 and P_2 are homogeneous at every instant during these processes.

[†] These numerical values were derived from a steady flow process governed by eqn. (15.24) and not from the type of cyclic process described. However, this equation will be derived directly from eqn. (6.1). Furthermore, the later discussion of § 15.8 will make clear that the previously defined method of the measurement of heat gives a value of C_u, whereas the actual measurements were of C_h. The difference is negligible to the quoted degree of accuracy.

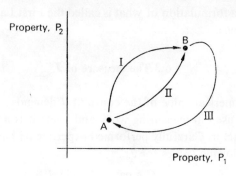

Fig. 6.3

For the first cyclic process, the application of eqn. (6.2) can be expressed

$$\left[\int_{A_\mathrm{I}}^{\uparrow B} W_\mathrm{to} + \int_{B_\mathrm{III}}^{\uparrow A} W_\mathrm{to}\right] = \left[\int_{A_\mathrm{I}}^{\uparrow B} Q_\mathrm{by} + \int_{B_\mathrm{III}}^{\uparrow A} Q_\mathrm{by}\right]$$

or, rewriting,

$$\int_{A_\mathrm{I}}^{\uparrow B} (W_\mathrm{to} - Q_\mathrm{by}) + \int_{B_\mathrm{III}}^{\uparrow A} (W_\mathrm{to} - Q_\mathrm{by}) = 0.$$

Similarly, for the second cyclic process,

$$\int_{A_\mathrm{II}}^{\uparrow B} (W_\mathrm{to} - Q_\mathrm{by}) + \int_{B_\mathrm{III}}^{\uparrow A} (W_\mathrm{to} - Q_\mathrm{by}) = 0.$$

Comparing this equation with the previous one results in

$$\int_{A_\mathrm{I}}^{\uparrow B} (W_\mathrm{to} - Q_\mathrm{by}) = \int_{A_\mathrm{II}}^{\uparrow B} (W_\mathrm{to} - Q_\mathrm{by}),$$

In other words, the quantity $\int_{A}^{\uparrow B} (W_\mathrm{to} - Q_\mathrm{by})$ is independent of the choice of the path of the process; it will depend only upon the two states A and B. Thus, as discussed in § 5.8, this quantity can be expressed as a change in a property. This property

THE FIRST LAW OF THERMODYNAMICS

is called the internal energy E, its definition being expressed as

$$E_B - E_A \equiv \int_A^B (W_{to} - Q_{by}). \qquad (6.3)$$

An origin from which E is measured is arbitrarily chosen, and a sign convention is implicit in this defining equation. Remembering also the discussion in § 5.8 on differentiability of a property, then, when this is possible,

$$E_B - E_A = \int_A^B dE.$$

A system can be split into many sub-systems, as is shown in the sketch of Fig. 6.4, where the sub-systems are numbered $1, 2, 3, \ldots, n$. Accepting that eqn. (6.3) applies to any system, it can be applied to each of these sub-systems in turn to give

$$\left. \begin{array}{l} W_{to_1} - Q_{by_1} = E_{B_1} - E_{A_1}, \\ W_{to_2} - Q_{by_2} = E_{B_2} - E_{A_2}, \\ \cdots\cdots\cdots\cdots\cdots\cdots \\ W_{to_n} - Q_{by_n} = E_{B_n} - E_{A_n}. \end{array} \right\} \qquad (6.4)$$

On adding all these equations together the first terms add up as

$$W_{to_1} + W_{to_2} + \ldots W_{to_n}.$$

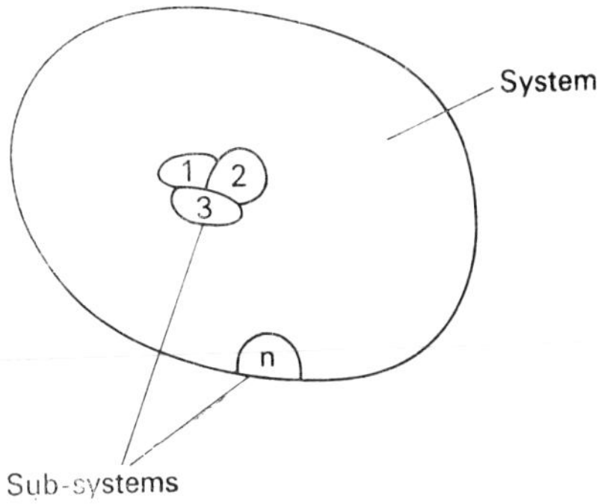

Fig. 6.4

As discussed in § 1.8, the work done to a system by a surface force has an equal and coincident reaction to its surroundings. Thus the amount of surface work for each sub-system cancels with those of its surrounding ones and when all these terms are added they cancel out with the exception of those applied at the boundary of the main system.

The total work done by the body forces is, by eqn. (5.11), the sum of that done to each sub-system. Thus the total work done to the whole system W_{to}, is given by

$$W_{to} = W_{to_1} + W_{to_2} + \ldots W_{to_n}. \tag{6.5}$$

The discussion of § 5.13 showed that heat also has an equal and coincident reaction. Thus, by the same reasoning, the total heat applied by the system Q_{by} is given by

$$Q_{by} = Q_{by_1} + Q_{by_2} + \ldots Q_{by_n} \tag{6.6}$$

and the right-hand side is the sum of the second terms of eqns. (6.4).

The change of internal energy of the whole system $(E_B - E_A)$ is given by eqn. (6.3), and so the summation of the eqns. (6.4), together with substitution of eqns. (6.5) and (6.6), gives

$$E_B - E_A = (E_{B_1} + E_{B_2} + \ldots E_{B_n}) - (E_{A_1} + E_{A_2} + \ldots E_{A_n}).$$

Or, the internal energy being measured from a fixed origin,

$$E = E_1 + E_2 + \ldots E_n,$$

and so it is an additive property.[†] Consequently, upon the discussion of § 5.9, there is value in defining an internal energy per mass unit which is given the symbol e.

---

† The limitation described in para. (b) of § 5.9 must be remembered here.

6.5 Work done by body forces

In the experiments performed to verify eqn. (6.1) and to derive the numerical value of J, the work was done to the system and the heat was applied by the system to its surroundings. It has already been demonstrated by eqn. (5.16) that

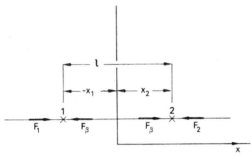

Fig. 6.5

$Q_{by} = -Q_{to}$. It is now possible to expand on the reservation that in general only the work by surface forces has an equal and coincident reaction to a surrounding system.

Consider the two systems containing electrical point charges of strength q_1 and q_2 and denoted 1 and 2 respectively in Fig. 6.5. The body force repelling them, F_β, is given by[1]

$$F_\beta = \frac{q_1 q_2}{\varepsilon l^2}, \tag{6.7}$$

where l is the distance between the charges and ε is the dielectric coefficient. Applying a force F_2 to restrain 2 in position means that no work is done to 2 by either F_2 or F_β. Thus by eqn. (6.3) the change in the internal energy of 2 is zero, or

$$\Delta E_2 = 0.$$

Applying a force F_1 to cause 1 to move between two positions of rest, means that there is no change in the kinetic energy of 1. Consequently, $\Delta E_1 = 0$, and also by eqn. (1.18) the work done by F_1, $W_{F_1 \text{ to}}$, is equal and opposite to that done by F_β, $W_{\beta_2 \text{ to}}$, or

$$W_{\beta_1 \text{ to}} = -W_{F_1 \text{ to}}. \qquad (6.8)$$

The value of $W_{F_1 \text{ to}}$ is not zero, for from eqns. (6.7) and (6.8), and with the notation of Fig. 6.5,

$$\begin{aligned}
W_{\beta_1 \text{ to}} &= -\int_{x_{1A}}^{x_{1B}} F \, dx_1 \\
&= -\frac{q_1 q_2}{\varepsilon} \int_{x_{1A}}^{x_{1B}} \frac{dx_1}{(x_2-x_1)^2} \\
&= -\frac{q_1 q_2}{\varepsilon} \left(\frac{1}{x_2-x_{1_B}} - \frac{1}{x_2-x_{1_A}} \right) \\
&= -W_{F_1 \text{ to}}.
\end{aligned}$$

Thus, as $W_{\beta_2 \text{ to}} = 0$, then the work done by the body force to 1 has not got an equal and coincident reaction upon the surrounding material system.

If systems 1 and 2 are now regarded as a single system, then for the process just described the change in internal energy, ΔE is

$$\Delta E = \Delta E_1 + \Delta E_2 = 0.$$

The work done to the system as 1 moves between x_{1_A} and x_{1_B} is $W_{F_1 \text{ to}}$ and to reconcile this finite work to the system with the zero internal energy change of the two sub-systems 1 and 2 so that eqn. (6.3) is still valid, it is necessary to assign an internal energy to the field of force and to think of the work as being required to change this field. As the dielectric coefficient still has a finite value when the charges are surrounded by a vacuum, then the internal energy of the force field can be associated with an immaterial system.

An alternative approach is only valid when interest is restricted to system 1 which moves in a stationary and conservative field of force obtained by keeping system 2 stationary. Then the discussion of § 1.9 applies and, using the reasoning of § 6.4, $W_{\beta_1 \text{to}}$ can be regarded as an energy which is associated with a specified position of system 1 and which is called its potential energy.† Thus in eqn. (6.3) the W_β term can be embodied in the $(E_B - E_A)$ term enabling W_{to} to be written as $-W_{\text{by}}$.[2] As also $Q_{\text{by}} = -Q_{\text{to}}$, then for this case eqn. (6.3) can be rewritten as

$$\int_A^B (Q_{\text{to}} - W_{\text{by}}) = E_B - E_A,$$

or, dropping the suffixes,

$$Q - W = \Delta E. \tag{6.9}$$

It is in this restricted form that the first law of thermodynamics is commonly quoted, and it will be so used in the later volumes without the suffixes to Q and W.

6.6 Application of electricity

The left-hand sketch of Fig. 6.6 illustrates a quantity of water containing an electrically conducting circuit. With the restrictions laid down in § 5.12, no work would be done to

---

† There is a danger of confusion in the use of this concept which must be guarded against and which does not arise in the previous idea of the energy of a field. If, for example, systems 1 and 2 are initially of zero internal energy and respectively at $x = -\infty$ and $x = +\infty$, then in bringing system 2 to $x = x_2$, say, no work is done because no field of system 1 exists as long as x_2 is finite. Thus at this position the internal energy of system 2 is still zero. As explained, when now system 1 is brought from infinity, its internal energy rises to a value of $(q_1 q_2)/[\epsilon(x_2 - x_1)]$. Thus the two identical systems in identical fields have different potential energies.

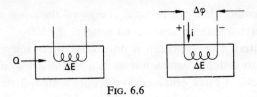

Fig. 6.6

this system during a process. A process in which heat of amount Q_{to} is applied would then result in a change of internal energy ΔE which is given by eqn. (6.9) as

$$Q_{to} = \Delta E. \tag{6.10}$$

As explained in § 5.1, when an electric current flows across a boundary, that boundary cannot enclose a system because the current flow is associated with a mass flow; this, for instance, is the essence of electro-deposition. But for many types of conductor, an example being a metal conductor, this transport cannot be observed on a continuum scale, and the contravention of the definition of a system is not apparent. Assuming that this approximation is valid for this example,† then the right-hand sketch of Fig. 6.6 illustrates the same system undergoing a different process in which continuous electric current is passed through the circuit. If additionally in this process thermal equilibrium is maintained between the system and its surroundings, then

$$Q_{to} = 0.$$

However, a change in state can be observed as a result of the passage of the electric current and, correspondingly, there is an internal energy change.

If the total electricity applied to the system N is measured by the expression

$$N \equiv \int i \, \Delta\varphi \, dt, \tag{6.11}$$

† When it is not, then the control volume method described in Chapter 15 is necessary.

THE FIRST LAW OF THERMODYNAMICS 77

where i is the current, $\Delta\varphi$ is the electrical potential difference or the electrical force field, and t is the time during the process, then for the same value of ΔE in the two processes, it is found that N is directly proportional to Q_{to}, the constant of proportionality being simply a units conversion factor. Making this factor unity provides a definition for the units of φ, the electrical potential. In the metre–kilogram–second–ampere system the corresponding unit of φ is the volt. Then, for these two processes,

$$N = Q_{to},$$

and so, from eqn. (6.10),

$$N = \Delta E. \qquad (6.12)$$

If the inlet current i_{in} and the outlet current i_{out} are not balanced, then a charge q given by

$$q = \int (i_{in} - i_{out})\, dt$$

is built up within this system. As described in § 6.5 this requires additional electricity, as defined by eqn. (6.11), to generate the resulting field of force.

A process in which heat is applied by and work and electricity are successively applied to a system can be expressed by eqns. (6.3) and (6.12) for each part as

$$-Q_{by} = (\Delta E)_1,$$
$$W_{to} = (\Delta E)_2,$$
$$N_{to} = (\Delta E)_3.$$

As the internal energy is an additive property, then

$$W_{to} - Q_{by} + N_{to} = (\Delta E)_1 + (\Delta E)_2 + (\Delta E)_3 = \Delta E. \quad (6.13)$$

As heat results from a lack of thermal equilibrium and work results from a lack of mechanical equilibrium so electricity results from a lack of electrical equilibrium as manifested by a finite value of the potential difference $\Delta\varphi$. Electricity, as the term is used here, is also transient in nature, acting only whilst a process is occurring.†

Inclusion of the further term N in eqn. (6.3) to represent the effects of electricity is usual practice in electrical engineering. Some writers in thermomechanics include electricity under the description of work by postulating an electric motor of an idealized kind.[3, 4] Discussion of this approach is excluded here but can be found in vol. 2 and the references just cited.

References

1. CONSTANT, F. W., *Fundamental Laws of Physics*, Addison-Wesley. Reading, Massachusetts, 1963, p. 202.
2. LANDSBERG, R., The thermodynamic system and the first law, *Bull. Mech. Eng. Educ.* **6**, (2), 185 (April–June 1967).
3. KEENAN, J. H., *Thermodynamics*, Wiley, New York, 1957, p. 2.
4. ZEMANSKY, M. W., *Heat and Thermodynamics*, 4th edn., McGraw-Hill, New York, 1957, p. 50.

---

† Electrical charge is the quantity that is stored.

CHAPTER 7

THE MANNER OF HEAT PROCESSES

7.1 The conduction of heat

A system, consisting of a small element of unstrained solid material of thickness δn and cross-sectional area δA, is illustrated in Fig. 7.1. If, for a period of time, a temperature difference

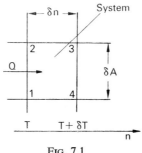

Fig. 7.1

δT exists between the boundaries 1–2 and 3–4, each boundary being at a uniform temperature, then heat Q would be applied to the boundary 1–2.

It could be experimentally observed that the heat rate $\dot{Q}$ is a function of the positive quantities δA, δT, and δn in the form

$$\dot{Q} \propto -\frac{\delta A\, \delta T}{\delta n}.$$

In the limit as $\delta n \to 0$ then also $\delta T \to 0$ and $\delta T/\delta n$ becomes the gradient of temperature $\partial T/\partial n$. Also, as $\delta A \to O$, $\dot{Q}/\delta A$ becomes the intensity of the heat rate η as defined by eqn. (5.18), so

$$\eta \propto -\frac{\partial T}{\partial n},$$

or, introducing a positive factor of proportionality k,

$$\eta = -k\frac{\partial T}{\partial n}. \tag{7.1}$$

In applying this relation it must be remembered that the temperature gradient is measured in a direction that is perpendicular to δA. The constant k is found to be a property and is called the coefficient of thermal conductivity.

When thermal diffusion is present, then sometimes this is incorporated in the relation defining the coefficient of thermal conductivity.[1]

In real processes of conduction of heat, for which the continuum assumption is valid, there is no discontinuity in temperature either within a system or at the boundary between it and the surroundings. If such a discontinuity existed then, for a thin slice of material enclosing it, the change in temperature across this slice would remain finite as its thickness was reduced to zero. Thus, from eqn. (7.1), the heat rate would be infinite, and no real process proceeds at an infinite rate.† When the density of a gas is very low so that the flow of it is in the slip flow region, then there is a discontinuity of temperature between the gas and the solid boundary.[3]

---

† But a mathematical model for the instant of commencement of a process which allows such a discontinuity can be useful.[2]

7.2 Temperature distribution along a bar

Figure 7.2 is a sketch of a bar of length X and cross-sectional area A which is undergoing a process devoid of work with $\dot{W} = 0$ at every instant during it. Specifying that the heat is acting only at the portions of the boundary indicated (1)–(1)

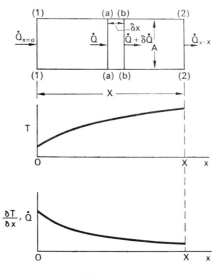

Fig. 7.2

and (2)–(2) in Fig. 7.2 would result in a variation of temperature within the bar in only the x-direction; such a distribution might be as sketched in the centre portion of this figure. This distribution is an instantaneous one; in general it will change with time. Correspondingly, at an instant there will be a variation of the temperature gradient $\partial T/\partial x$ and hence, from eqn. (7.1), also of $\dot{Q}$; the lower portion of Fig. 7.2 suggests such variations. Thus $\dot{Q}$ is a function not only of time as discussed in § 5.14, but also in this case it is a function of x.

Temperature discontinuities have been excluded, but in addition discontinuities in $\partial T/\partial x$ at any position within the bar can be exluded for the following reasons. In §5.14 it was demonstrated that $\dot{Q}$ has an equal and coincident reaction: any station along the bar can be regarded as a portion of a system boundary, and approaching this boundary from either side thus results in a unique value of $\dot{Q}$; there are no discontinuities in its variation. As the state at any station is single-valued so also are the properties there, and in particular the coefficient of thermal conductivity is single-valued. It thus follows from eqn. (7.1) that so also is $\partial T/\partial x$.

Consider a thin slice of the bar as a system. Such a slice is marked (a)–(a), (b)–(b) in Fig. 7.2. Calling $\dot{Q}$ the rate of heat to this system across section (a)–(a) then, assuming now and justifying later that $\dot{Q}$ is differentiable, the rate of heat applied by the slice across (b)–(b) can be written

$$\dot{Q} + \delta\dot{Q} = \dot{Q} + \frac{\partial \dot{Q}}{\partial x}\,\delta x.$$

Thus the net rate of application of heat to the slice is

$$-\delta\dot{Q} = -\frac{\partial \dot{Q}}{\partial x}\,\delta x.$$

In time δt the heat applied is

$$-\frac{\partial \dot{Q}}{\partial x}\,\delta x\,\delta t. \tag{7.2}$$

For this case eqn. (7.1) can be written

$$\dot{Q} = -kA\frac{\partial T}{\partial x},$$

and, differentiating and substituting into eqn. (7.2), gives the heat applied to the system as

$$A\frac{\partial}{\partial x}\left(k\frac{\partial T}{\partial x}\right)\delta x\,\delta t. \tag{7.3}$$

The increase in the internal energy of the slice during this time can be written

$$\delta m \frac{\partial e}{\partial t} \delta t, \qquad (7.4)$$

where the mass of the slice δm is given by

$$\delta m = \rho A\, \delta x. \qquad (7.5)$$

Performance of the experiments described in § 6.1 led to the formulation of the first law of thermodynamics which compared the states at each end of a process, both being of uniform temperature and of no relative motion within the system. The present case of a solid bar satisfies only the second of these conditions during the process. But by taking the slice of the bar of thickness δx as a system, then in the limit as $\delta x \to 0$ the temperature difference across the slice will also tend to zero and then the first of these two conditions is also satisfied. So applying eqn. (6.9) by equating eqns. (7.3) and (7.4) and substituting eqn. (7.5) results in

$$A \frac{\partial}{\partial x}\left(k \frac{\partial T}{\partial x}\right) \delta x\, \delta t = \rho A\, \delta x \frac{\partial e}{\partial t} \delta t$$

or

$$\frac{\partial}{\partial x}\left(k \frac{\partial T}{\partial x}\right) = \rho \frac{\partial e}{\partial t}. \qquad (7.6)$$

This differential equation provides a general relation between the temperature distribution at any instant and the rate at which the internal energy is changing at any point. The assumption that $\dot{Q}$ is differentiable can now be justified. If an approach to a section of the bar from both sides gave two values of $\partial \dot{Q}/\partial x$, then from eqn. (7.6) that position of the bar would have two values of $\partial e/\partial t$. The real rate of change of the state of a system can have only a single value. Thus there can be no discontinuities in $\partial \dot{Q}/\partial x$.

In the special case when this heat process in the bar has settled down and is proceeding at a constant rate, then at any position along the bar there will be no change in the local state with respect to time. Thus

$$\frac{\partial e}{\partial t} = 0,$$

and then, from eqn. (7.6),

$$\frac{\partial}{\partial x}\left(k\,\frac{\partial T}{\partial x}\right) = 0.$$

Integrating this once gives

$$k\,\frac{\partial T}{\partial x} + C = 0, \qquad (7.7)$$

where C is the constant of integration which, from eqn. (7.1), is equal to η.

The value of k for solids can vary significantly with temperature,[4] the form of this variation differing greatly between materials. For most purposes it can be taken as a function of only temperature so that eqn. (7.7) can be directly integrated to give

$$\int k\,dT + Cx + D = 0,$$

where D is another constant of integration.

Values of $\int_0^T k\,dT$ for three materials are shown plotted in Fig. 7.3. It can be seen that for many purposes a linear approximation to this integral would suffice. But when temperature differences are small, then k can more accurately be taken to have the constant value corresponding to the mean temperature. Then the above equation becomes

$$kT + Cx + D = 0,$$

and the temperature variation is linear.

When valid, this latter relation has a useful application to a steady heat process applied across a wall. If, as sketched in

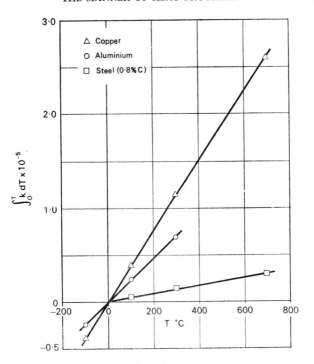

Fig. 7.3

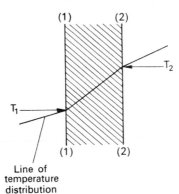

Fig. 7.4

Fig. 7.4, the two temperatures on the surfaces of the wall can be specified, then $\partial T/\partial x$ is given by the ratio of the difference between these two temperatures to the wall thickness. Substituting into eqn. (7.1) enables the value of $\dot{Q}$ to be calculated.

7.3 The radiation of heat through a vacuum

The final example given in § 3.1 introduced heat by radiation. This manner of applying heat can occur across a vacuum, and is realized by photons which are "particles" of zero mass but of finite energy e_p and of finite momentum.

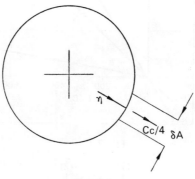

Fig. 7.5

The photons move at the velocity of light in *vacuo* c and in random directions. Because of this they cross from the side facing the radiating object, a unit area of any plane drawn in space at a rate of $\frac{1}{4}Cc$, where C is their concentration in number per unit volume.[5]

If radiation is taking place from a sphere as illustrated in Fig. 7.5, then, by eqn. (5.19), the heat applied by the sphere to its surroundings across an area δA in time δt is $\eta \, \delta A \, \delta t$. Regarding these surroundings as a system, application of eqn. (6.9)

gives
$$\eta\,\delta A\,\delta t = \Delta E.$$

Across a plane adjacent to the surface the number of photons that has passed to the surroundings in this time is
$$\tfrac{1}{4}Cc\,\delta A\,\delta t.$$

The energy carried by these photons, which will be ΔE, is thus
$$\tfrac{1}{4}Cce_p\,\delta A\,\delta t = \Delta E,$$
and so
$$\eta = \tfrac{1}{4}Cce_p. \tag{7.8}$$

The energy per unit volume E' will have a value at the surface of Ce_p, and so eqn. (7.8) can be written
$$E' = \frac{4\eta}{c}. \tag{7.9}$$

As with force fields, consistency with the first law of thermodynamics requires the attribution of an internal energy to an immaterial system which now is the vacuum occupied by photons having no mass.

Because photons also possess momentum, their collision with a solid surface imparts a pressure to it. With molecules which reflect from a solid surface in the manner of perfectly elastic solids, the kinetic theory gives the resulting pressure as[6]
$$p = \tfrac{1}{3}\varrho\bar{c}_m^2,$$

where $\bar{c}_m^2$ is the mean square molecular speed. As the energy per unit volume of this translational speed E' is $\tfrac{1}{2}\varrho\bar{c}_m^2$, then, for molecules,
$$p = \tfrac{2}{3}E'.$$

But when photons are absorbed and not reflected by a surface, then the momentum change is half of the above value and thus
$$p = \tfrac{1}{3}E'. \tag{7.10}$$

It is observed that, for radiation, η is a function of only the temperature of the emitting surface and thus, from eqn. (7.9),

$$E' = E'(T). \qquad (7.11)$$

A general relation, whose proof is now anticipated,[7] for gases, has the form

$$\left(\frac{\partial E}{\partial V}\right)_T = T\left(\frac{\partial p}{\partial T}\right)_V - p. \qquad (7.12)$$

Substitution from eqn. (7.10), noting eqn. (7.11) and that $E = E'V$, gives

$$E' = T\frac{1}{3}\frac{dE'}{dT} - \frac{1}{3}E'$$

or

$$\frac{dE'}{E'} = 4\frac{dT}{T}.$$

Integration results in

$$\log E' = 4\log T + \log K,$$

where $\log K$ is the constant of integration. Thus

$$E' = KT^4$$

and then, from eqn. (7.9),

$$\eta = \frac{Kc}{4}T^4, \qquad (7.13)$$

a result known as the Stefan law.

This relation has been derived assuming that all the photons are absorbed by a receiving surface. Such a surface is known as a black body surface. When only the proportion α is absorbed, then the rate of heat absorbed $\dot{Q}^{(\alpha)}$ is related to that incident upon the surface $\dot{Q}^{(i)}$ by,

$$\dot{Q}^{(\alpha)} = \alpha \dot{Q}^{(i)}. \qquad (7.14)$$

Emission, like absorption, depends upon the nature of a surface. The relation between the rate of emission of heat from a surface $\dot{Q}^{(e)}$ to that from a black body surface $\dot{Q}_b^{(e)}$ is written

$$\dot{Q}^{(e)} = \varepsilon \dot{Q}_b^{(e)}. \tag{7.15}$$

The upper sketch (a) of Fig. 7.6 is of a sphere with a black-body surface that is separated by a cavity from an enclosing contain-

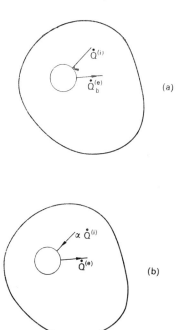

FIG. 7.6

er. If these surfaces are in thermal equilibrium then there is no net heat between them and so the heat rate incident on the centre sphere $\dot{Q}^{(i)}$ must exactly counterbalance that emitted from its black-body surface $\dot{Q}_b^{(e)}$, or

$$\dot{Q}^{(i)} = \dot{Q}_b^{(e)}. \tag{7.16}$$

If, as shown in the second sketch (b), of Fig. 7.6, the inner sphere is replaced by another of the same size which has not a black-body surface, then its absorption is $\alpha \dot{Q}^{(i)}$. For balance this must equal its emission $\dot{Q}^{(e)}$ or

$$\alpha \dot{Q}^{(i)} = \dot{Q}^{(e)}. \tag{7.17}$$

Comparison of eqns. (7.16) and (7.17) gives

$$\dot{Q}^{(e)}/\dot{Q}_b^{(e)} = \alpha.$$

Then, from eqn. (7.15),

$$\varepsilon = \alpha. \tag{7.18}$$

Thus the emission is a maximum for a black-body, for then $\varepsilon = 1$. In this case eqn. (7.13) is written

$$\eta = \sigma T^4, \tag{7.19}$$

and consequently for other surfaces

$$\eta = \sigma \varepsilon T^4. \tag{7.20}$$

The coefficient σ is a universal coefficient having the value $5 \cdot 67 \times 10^{-8}$ kg sec^{-3} °K^{-4}. The values of ε and α depend upon the texture of the surface and upon its temperature. Values are tabulated[8] showing the uncertainty in estimating them. Whilst eqns. (7.19) and (7.20) have been derived for systems in equilibrium, they are used generally for radiation heat processes.

In the example just discussed, all the radiation from the inner sphere falls upon the walls of the container; only some of that from the container falls on the sphere, the rest falling instead on another part of the container. The proportion incident upon the sphere is measured by a geometric view factor F. If the object within the cavity is not a sphere but a shape containing a concavity, then only some of its emission, as measured by another F factor, would travel straight to its surroundings; the rest would be intercepted by its own surface. If the intercepting surfaces are not black-body ones, then multiple reflections of the radiation take place.

In the case of two black bodies, denoted respectively 1 and 2, and having areas respectively of A_1 and A_2 and temperatures of T_1 and T_2, the emitted heat rates are, respectively,

$$\dot{Q}_1^{(e)} = \sigma A_1 T_1^4,$$
$$\dot{Q}_2^{(e)} = \sigma A_2 T_2^4.$$

If these bodies are isolated from all material surroundings, then the incident heat rates are, respectively,

$$\dot{Q}_1^{(i)} = F_{21}\sigma A_2 T_2^4,$$
$$\dot{Q}_2^{(i)} = F_{12}\sigma A_1 T_1^4,$$

where F_{21} and F_{12} are the respective view factors. When there is thermal equilibrium between two bodies, then $T_1 = T_2$ and

$$\dot{Q}_1^{(i)} = \dot{Q}_2^{(i)}$$

so that $\qquad F_{12}A_1 = F_{21}A_2.$ \hfill (7.21)

This result being a geometrical one it is also valid when equilibrium does not exist. Values of F have been computed for simple shapes.[9] Then the net rate of heat by 1 is

$$\dot{Q}_{\text{by} 1} = \dot{Q}_1^{(e)} - \dot{Q}_1^{(i)}$$
$$= \sigma A_1 (T_1^4 - F_{12} T_2^4)$$

and that to 2 is

$$\dot{Q}_{\text{to} 2} = \dot{Q}_2^{(i)} - \dot{Q}_2^{(e)}$$
$$= \sigma A_1 F_{12} \left(T_1^4 - \frac{T_2^4}{F_{21}} \right).$$

As the view factors F have values lying in the range $0 < F \leq +1\cdot 0$, then these two heat rates can never be equal for any real value of T_2/T_1. This illustrates that care must be taken in the computation of the radiation heat for use in the first law of thermodynamics.

A further example is given by the radiation from a body denoted by 1, enclosed in a black-body cavity denoted by 2.

The previous relations are now

$$\dot{Q}_1^{(e)} = \varepsilon_1 \sigma A_1 T_1^4,$$
$$\dot{Q}_1^{(i)} = \alpha_1 F_{21} \sigma A_2 T_2^4,$$

and $\qquad A_1 = F_{21} A_2.$

Noting eqn. (7.18), the rate of heat from 1 is

$$\begin{aligned}\dot{Q}_{\text{by1}} &= \varepsilon_1 \sigma A_1 T_1^4 - \alpha_1 F_{21} \sigma A_2 T_2^4 \\ &= \sigma A_1 (\varepsilon_1 T_1^4 - \alpha_1 T_2^4) \\ &= \varepsilon_1 \sigma A_1 (T_1^4 - T_2^4).\end{aligned}$$

A common feature of the relative magnitudes of $\dot{Q}_1^{(e)}$ and $\dot{Q}_1^{(i)}$ is illustrated by a numerical example. An electric fire receiving 1 kW of electricity has an emissivity of about 0·8. Its temperature, which is limited by the mechanical strength of the wire element, is of an orange colour which is roughly equivalent to 1400° K. Situated in surroundings at 290° K, the surface area needed is

$$\begin{aligned}A_1 &= \frac{10^3}{0\cdot 8 \times 5\cdot 75 \times 10^{-8}(1\cdot 4^4 \times 10^{12} - 2\cdot 9^4 \times 10^4)} \\ &= 5\cdot 7 \times 10^{-3} \text{ m}^2 \\ &= 8\cdot 8 \text{ in}^2.\end{aligned}$$

It is seen that, to the order of accuracy possible, the term T_2^4 is quite negligible compared with T_1^4.

7.4 The radiation of heat through an opaque medium

Clean gases are almost transparent to heat radiation; liquids generally are not so. The latter will absorb radiation at a rate $\mathcal{A}$ per unit volume and will also emit it at a rate $\mathcal{E}$ per unit volume.

An element of an absorbing medium is sketched in Fig. 7.7. A simple example is provided by a parallel beam of radiation in the x-direction with the temperature of the medium so low that by eqn. (7.20) its self-emission is negligible. The discussion is now closely similar to that of § 7.2. As marked in Fig. 7.7, the rate of passage of radiation at the left is $\eta \delta y$. In passing through the element, radiation will be absorbed at a rate

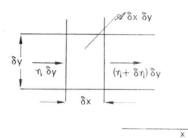

Fig. 7.7

$\mathcal{A} \, \delta n \, \delta y$. It will then leave the element at a rate $(\eta + \delta \eta)$. For balance,[†]

$$\eta \, \delta y - (\eta + \delta \eta) \, \delta y = \mathcal{A} \, \delta x \, \delta y. \tag{7.22}$$

Now

$$\delta \eta = \frac{\partial \eta}{\partial x} \, \delta x,$$

and it is found that $\mathcal{A}$ is proportional to η as expressed by

$$\mathcal{A} = c_e \eta,$$

where c_e is called the extinction coefficient. Thus

$$\frac{\partial \eta}{\partial x} + c_e \eta = 0.$$

The solution of this is

$$\eta = \eta_0 e^{-c_e x},$$

where η_0 is the value of the radiation heating rate at $x = 0$

[†] To be precise, this is the equation that defines $\mathcal{A}$.

By this equation, the radiation rate falls exponentially and has half its initial value when

$$e^{-c_e x} = 0.5$$

or

$$x = \frac{0.693}{c_e}.$$

Commonly in engineering problems, heating occurs simultaneously by radiation and conduction. As explained in § 5.13, these two components are additive. Where this combination occurs in a fluid in motion, the analysis can become complex.[10]

7.5 The rate of a process

Equating eqns. (7.2) and (7.4) gives

$$-\frac{\partial \dot{Q}}{\partial x} \delta x = \frac{\partial e}{\partial t} \delta m$$

$$= \frac{\partial}{\partial t} (e\, \delta m). \qquad (7.23)$$

As this relation is valid for each and every elemental slice along the bar then it can be summed for them all giving,

$$\int_0^x -\frac{\partial \dot{Q}}{\partial x} \delta x = \int_0^x \frac{\partial}{\partial t}(e\, \delta m).$$

Therefore

$$-\int_0^x d\dot{Q} = \frac{\partial}{\partial t} \int_0^x e\, dm.$$

The integral on the left-hand side of this equation becomes $\dot{Q}$ applied to the bar at $x = 0$, denoted $\dot{Q}_{x=0}$ in Fig. 7.2, minus that applied by the bar at $x = X$ similarly denoted $\dot{Q}_{x=x}$, and this difference is the total $\dot{Q}$ applied to the bar. The integral on the right-hand side is the total internal energy of the bar E, so that this equation becomes

$$\dot{Q} = \frac{\partial E}{\partial t}. \tag{7.24}$$

Whilst the first law of thermodynamics, as expressed by eqn. (6.9), gives a relation for the overall change resulting from a process, eqn. (7.24) gives a relation for the rate with which the process is proceeding. Though eqn. (7.24) was derived assuming a one-dimensional variation of temperature with only x, the analysis can readily be extended to show that the same result is obtained for a general three-dimensional heat process.

A comparison of eqns. (7.22) and (7.23) shows that

$$\mathcal{A} = \frac{\partial}{\partial t}(\varrho e),$$

so that the rate equation also applies to heat by radiation.

Both terms in eqn. (7.24) being rate terms, the integration can be performed with respect to time between limits t_1 and t_2, say, giving

$$\int_{t_1}^{t_2} \dot{Q}\, dt = \int_{t_1}^{t_2} \frac{\partial E}{\partial t}\, dt.$$

Therefore

$$\left.\vphantom{\int}\right|_{t_1}^{t_2} Q = E_{t_2} - E_{t_1},$$

and thus the first law of thermodynamics for this heat process is now extended to apply between any two times during a process even though there is not uniformity of state throughout the system at these times.

7.6 The directional nature of heat by conduction

Figure 7.8 illustrates a small triangular portion of material, denoted ABC, bounded by two lines along which the temperature has a constant value of respectively T and of $T+\delta T$.

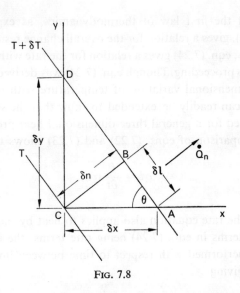

Fig. 7.8

The distance between these two lines being δn, as shown, then the temperature gradient is $\partial T/\partial n$. Thus from eqn. (7.1) the rate of application of heat $\dot{Q}_n$ across the line AB of length δl and per unit thickness, perpendicular to the x–y plane, is

$$\dot{Q}_n = -k \frac{\partial T}{\partial n} \delta l.$$

Referring to Fig. 7.8, it is seen that

$$\cot \theta = \frac{\delta l}{\delta n} = \frac{\delta x}{\delta y},$$

so that

$$\dot{Q}_n = -k\frac{\partial T}{\partial y}\,\delta x. \qquad (7.25)$$

Here, $\partial T/\partial y$ is the component, perpendicular to the boundary AC, of the temperature gradient.

The zero'th law of thermodynamics refers to the consequence of a difference in temperature existing between two systems before the commencement of a process and does not directly give any information during a process when, from the discussion of § 7.1, there is no discontinuity of temperature at a system boundary. So it is now assumed that $\dot{Q} = 0$ across the line BC and this assumption will be justified later.

Then, for a steady-state process on the triangular element, eqn. (7.24) gives

$$-\dot{Q}_n + \dot{Q}_y = 0,$$

where $\dot{Q}_y$ is the rate of application of heat to the element across the line AC and per unit thickness. Thus, substituting from eqn. (7.25),

$$\dot{Q}_y = -k\frac{\partial T}{\partial y}\,\delta x. \qquad (7.26)$$

Now the line AC is of length δx, and so this expression shows that when the heat is to be evaluated along a system boundary which is not at a uniform temperature, the length of the boundary must be multiplied by the component of the temperature gradient normal to the boundary, to give the rate of heat intensity.

This result is independent of the value of θ in Fig. 7.8. In particular it is valid when $\theta = \pi/2$, and then the component of the temperature gradient perpendicular to the boundary is zero and so the heat rate by eqn. (7.26) is zero. This then justifies the prior assumption that the heat rate across the line BC is zero.

Heat rate is not a vector, but the intensity of the rate of heat η is.[12]

This is readily seen by first noting that analogous to eqn. (7.26),

$$\dot{Q}_x = -k \frac{\partial T}{\partial x} \delta y,$$

and so the total heat rate $\dot{Q}$ applied along the line AD in Fig. 7.8 is

$$\dot{Q} = \dot{Q}_x + \dot{Q}_y.$$

Noting that the length of AD is $\delta y/\sin \theta$, then substitution from eqn. (7.1), gives

$$\frac{\eta \, \delta y}{\sin \theta} = \eta_x \, \delta y + \eta_y \, \delta x,$$

and thus

$$\eta = \eta_x \sin \theta + \eta_y \cos \theta.$$

This shows that η is a vector having the components η_x and η_y and a direction to the x-axis of $(\pi/2 - \theta)$.

References

1. BIRD, R. B., CURTISS, C. F. and HIRSCHFELDER, J. O., *Fluid Mechanics and the Transport Phenomena*, Chem. Eng. Prog. Symp., Series No. 16, vol. 51, 1955, p. 77.
2. CARSLAW, H. S. and JAEGER, J. C., *Conduction of Heat in Solids*, 2nd edn., Oxford, 1959, pp. 18, 476.
3. PATTERSON, G. N., *Molecular Flow of Gases*, Wiley, New York, 1956, p. 187.
4. KAYE, G. W. C. and LABY, T. H., *Tables of Physical and Chemical Constants and some Mathematical Functions*, 12th edn., Longmans, London, 1960, pp. 51, 52.
5. JEANS, J. H., *The Dynamical Theory of Gases*, 3rd edn., Cambridge, 1921, p. 121.
6. *Ibid.*, p. 116.
7. ZEMANSKY, M. W., *Heat and Thermodynamics*, 4th edn., McGraw-Hill, New York, 1957, p. 249.

8. Bird, R. B., Stewart, W. E., and Lightfoot, E. N., *Transport Phenomena*, Wiley, New York, 1960, p. 432.
9. McAdams, W. H., *Heat Transmission*, 3rd edn., McGraw-Hill, New York, 1954, p. 68.
10. Chang, Y. P., A potential treatment of energy transfer by conduction, radiation and convection, *AIAA Jl.* **5** (5), 1024 (May 1967).
11. Tsidil'kovskii, I. M., *Thermomagnetic Effects in Semiconductors*, Infosearch, London, 1962 (tr. A. Tybulewicz), p. 3.
12. Carslaw, H. S. and Jaeger, J. C., *Conduction of Heat in Solids*, 2nd edn., Oxford, 1959, § 1.3.

CHAPTER 8

APPLICATION OF THE FIRST LAW OF THERMODYNAMICS TO SOLIDS

8.1 The sliding friction process

Figure 8.1 illustrates an experiment in which two identical homogeneous blocks are in contact. If both blocks together form a system which at all times is kept in thermal equilibrium with the surroundings by adjustment of the temperature of the

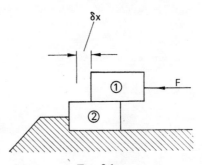

Fig. 8.1

latter, then, during a process in which the upper block under the application of a force F moves a distance δx,

$$Q = 0 \quad \text{and} \quad -W = +W_{\text{to}} = F\,\delta x.$$

Applying the first law of thermodynamics, eqn. (6.9) gives the resulting internal energy increase ΔE as

$$\Delta E = F\,\delta x.$$

THE FIRST LAW OF THERMODYNAMICS TO SOLIDS

Before and after this process the properties will be uniform throughout the system. So, as the internal energy is additive, the change in internal energy of each block will be $\frac{1}{2}\Delta E$.

This process can also be studied by considering the stationary block, indicated by 2 in Fig. 8.1, as a system. During the process there is no temperature difference between the two blocks at their surface of contact,[†] and so for block 2, $Q = 0$.

But as the change in internal energy is $\Delta E/2$, then

$$W_{\text{to}} = \tfrac{1}{2}\Delta E$$
$$= \tfrac{1}{2} F\,\delta x.$$

But the force at the surface of contact is F and so the equivalent movement of this surface is $\frac{1}{2}\,\delta x$. As this work has an equal and opposite reaction upon the upper block the total work done to the latter is

$$W_{\text{to}} = F\,\delta x - \tfrac{1}{2} F\,\delta x$$
$$= \tfrac{1}{2} F\,\delta x.$$

By considering block 1 as a system this result is seen to be consistent with the already determined internal energy change of $\frac{1}{2}\Delta E$ for this block.

This mean movement of the contact surface is in accord with the present knowledge of rubbing. Either projections are in contact, thereby being broken off sideways, or the material melts and a fluid lubrication phenomena occurs. For both cases, on the average the dividing surface moves half the distance through which the upper block is traversed. The usual assumption of slip, made in mechanical studies, is not valid here.

[†] The existence of temperature variations within each block during the process does not invalidate this equality. By symmetry, across the contact surface the temperature gradient is zero, and so by eqn. (7.1) the heat is zero.

8.2 The constant temperature stressing process

When a load is applied to a solid, a stress distribution is developed within it. A simple example of this is that of a rod, of constant cross-sectional shape, and area A, under only a tensile force. This is illustrated in Fig. 8.2. Sufficiently far

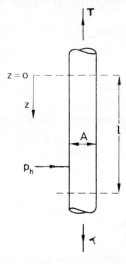

Fig. 8.2

from the end supports the stress is uniform throughout a length l of the rod and is in the axial direction. The rod being in contact with a large reservoir of internal energy at a constant temperature, such as the atmosphere, then, when the tensile force **T** is changed slowly, a process will take place in which the rod remains effectively at a constant temperature T. During this process the length of the rod l will change. The rate of change of length with respect to the force at this constant temperature is denoted by $(\partial l/\partial \mathbf{T})_T$. An associated quantity, known as

THE FIRST LAW OF THERMODYNAMICS TO SOLIDS

Young's modulus Y, is defined by the relation

$$Y \equiv \frac{l_0}{A_0}\left(\frac{\partial \mathbf{T}}{\partial l}\right)_T, \tag{8.1}$$

where suffixes 0 indicate values corresponding to an unstressed condition. Excluding the effects of creep and of plastic flow beyond the yield point, in this slow process Y is found to be a constant. It varies slightly with the temperature[†] of the process and is a property of the material. Thus from eqn. (8.1), $\mathbf{T}$ varies linearly with l, integration giving

$$\mathbf{T} = YA_0\left(\frac{l-l_0}{l_0}\right). \tag{8.2}$$

In stretching the rod through a short distance δl, work is done to the rod by the tensile force of amount $\mathbf{T}\,\delta l$. When eqn. (8.2) is applicable, $\mathbf{T}$ is a single valued function of l and so the work during this constant temperature process W_T can be evaluated by

$$W_T = \int_{l_0}^{l} \mathbf{T}\, dl. \tag{8.3}$$

From eqn. (8.1)

$$dl = \frac{l_0}{YA_0}\, d\mathbf{T}.$$

Therefore,

$$W_T = \frac{l_0}{YA_0}\int_0^{\mathbf{T}} \mathbf{T}\, d\mathbf{T}$$

$$= \frac{l_0 \mathbf{T}^2}{2YA_0} \tag{8.4}$$

[†] An alloy of nickel, steel, and chromium has been developed for which the variation of Y with temperature is very small indeed. This material is used for the balance springs of watches.

104 THERMOMECHANICS

It can be seen that because **T** is a single-valued function of $l^†$ then this constant temperature stretching process can be reversed and the final state will be identical with the initial one. This is called a reversible process. As will be shown later the process need not be very slow for it to be reversible.‡

8.3 The effect of hydrostatic pressure

In a real process **T** is not the only force acting upon the solid. The surroundings, such as the atmosphere or a liquid in which the solid might be immersed, exert a pressure force and this

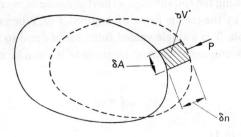

FIG. 8.3

together with body forces such as gravity can do work in a deformation. To evaluate this work consider the solid illustrated in its unstressed state by a solid line contour in Fig. 8.3 and in its stressed state by the dotted line contour.

The numerical example in § 11.7 shows that during the deformation the tangential stress exerted by the surrounding fluid is quite negligible compared to the normal stress which will be uniform. Hence eqn. (2.6) can be substituted into eqn. (5.13)

---

† Because creep and plastic flow have been excluded, hysteresis effects are not present.[1]

‡ If, at the other extreme, it is very fast, Y may not be constant during it.

THE FIRST LAW OF THERMODYNAMICS TO SOLIDS   105

to give the pressure work W_p as

$$W_p = \int \sigma \infty V$$
$$= -p \int \infty V$$
$$= -p \, \Delta V, \quad (8.5)$$

where ΔV is the total volume increase of the solid that results from the stress application.

8.4 Work done by gravity

To evaluate the work done by gravity suppose a rod, which is under tension, to be supported in a vertical position from its upper end. If z be the distance along the rod measured downwards from this end as shown in Fig. 8.2, then, as a result of stressing, an element moves vertically a distance which, for uniformity of stress and from eqn. (8.2), is, by simple proportion,

$$z \frac{\Delta l}{l_0},$$

and this is equal to $q_\beta \, \delta t$. Here Δl is written for $(l - l_0)$. The body force per unit mass β is, from eqn. (1.11), g, and so the work W_β is given by the second term of eqn. (5.12) as

$$W_\beta = \int g \frac{z}{l_0} \Delta l \, \delta m.$$

The mass of the element δm is given by

$$\delta m = \varrho A \, \delta z,$$

and so
$$W_\beta = g\varrho A \frac{\Delta l}{l_0} \int_0^{l_0} z \, dz$$

$$= \frac{1}{2} g\varrho A l_0 \, \Delta l. \quad (8.6)$$

The quantity $g_0 A l_0$ is the weight of the rod and $\Delta l/2$ is the movement of its centre of gravity, and so this result is one that might be expected.

A numerical example now provides a comparison of the relative values of these three work terms. This example is of a stress $\mathbf{T}/A_0$ of $1\cdot 5 \times 10^4$ lb in^{-2} acting upon a mild steel rod whose unstrained length and diameter are respectively 10 in. and 1·0 in. For mild steel $Y = 3 \times 10^7$ lb in^{-2} and so, from eqn. (8.4),

$$W_\mathbf{T} = \frac{10(1\cdot 5 \times 10^4)^2 \pi}{2 \times 3 \times 10^7 \times 4} = 29\cdot 4 \text{ lb in.}$$

To evaluate the work done by the hydrostatic pressure, the increase in volume ΔV must be known. When a bar is strained lengthwise by an amount $\Delta l/l_0$ there is a corresponding proportional change in the radius of the bar of an amount $\Delta r/r_0$. The ratio of these two strains is called Poisson's ratio, which is denoted by P_0, and so

$$\frac{\Delta r}{r_0} \equiv -P_0 \frac{\Delta l}{l_0}, \tag{8.7}$$

the negative sign indicating that an increase in length results in a decrease of radius, the property P_0 being positive. P_0 is effectively a constant. The change in volume then is given by

$$\frac{V_0 + \Delta V}{V_0} = \frac{(l_0 + \Delta l)\pi(r_0 + \delta r)^2}{l_0 \pi r^2},$$

which with substitution of eqn. (8.7) and for $\Delta l/l_0 \ll 1$, leads to

$$\frac{\Delta V}{V_0} = (1 - 2P_0)\frac{\Delta l}{l_0}. \tag{8.8}$$

For this example, eqn. (8.2) gives

$$\frac{\Delta l}{l_0} = \frac{\mathbf{T}}{YA_0} = \frac{1\cdot 5 \times 10^4}{3 \times 10^7} = 0\cdot 5 \times 10^{-3},$$

and so, as $P_0 = 0.29$ for steel,

$$\Delta V = \frac{10\pi}{4} 0.5 \times 10^{-3}(1 - 2 \times 0.29)$$

$$= 1.65 \times 10^{-3} \text{ in}^3.$$

Then, if the hydrostatic pressure is atmospheric pressure having the value 14·7 lb in^{-2}, the work is, by eqn. (8.5), given by

$$W_p = -14.7 \times 1.65 \times 10^{-3} = 0.024 \text{ lb in}.$$

The density of steel is 490 lb ft^{-3} and so[†]

$$\varrho = \frac{490}{32.2 \times 12^3} = 0.89 \times 10^{-2} \text{ slugs in}^{-3},$$

and thus, from eqn. (8.6),

$$W_\beta = \frac{1}{2} 32.2 \times 0.89 \times 10^{-2} \times \frac{\pi}{4} \times 10^2 \times 0.5 \times 10^{-3}$$

$$= 0.0056 \text{ lb in}.$$

Usually, then, the work done by the pressure and by gravity is negligible compared with that done by the tensile force. But this is a case where general conclusions must not be drawn from a special example. Though the conclusion is true in this case for a typical stress of the value met with in engineering work, it will be noticed from a comparison of eqns. (8.4) and (8.6) and using eqn. (8.2), that

$$\frac{W_\beta}{W_T} = \frac{F_g}{\mathbf{T}},$$

where F_g is the weight force. This ratio tends to infinity as

---

[†] This is an illustration of the practice adopted in this book whereby all data is converted into a common system of units—not always the same system for all problems—before substitution into equations.

$T \to 0$. Similarly, eqns. (8.2), (8.5), and (8.8) lead to the result

$$\frac{W_p}{W_T} = -\frac{2pA_0(1-2P_0)}{T},$$

which again tends to ∞ as $T \to 0$.[†]

8.5 The heat process

When the temperature of a solid is changed a corresponding change in volume V takes place. A coefficient of volume thermal expansion β is defined by

$$\beta \equiv \frac{1}{V_0}\left(\frac{\partial V}{\partial T}\right)_T, \quad (8.9)$$

where the differential coefficient is formed under conditions of constant applied stress.

If the material is homogeneous, then when the change in volume takes place there is no change in the shape of the solid. If a typical dimension of the solid which describes its size is l_0 then a coefficient of linear thermal expansion α is defined by,

$$\alpha \equiv \frac{1}{l_0}\left(\frac{\partial l}{\partial T}\right)_T. \quad (8.10)$$

The volume is proportional to the cube of l and so originally

$$V_0 \propto l_0^3, \quad (8.11)$$

and after a small expansion during which V_0 changes to $V_0 + \delta V$ and l_0 changes to $l_0 + \delta l$, then,

$$V_0 + \delta V \propto (l_0 + \delta l)^3.$$

The right-hand side of this latter relation can be expanded by

---

[†] Equally it must not be assumed that body and pressure forces are insignificant in the design of a structure. Most of the stress in a long-span bridge is due to its weight and the wind loads that act. Other examples abound.

the Binomial theorem. Doing this and retaining only first-order terms gives

$$V_0 + \delta V \propto l_0^3 \left(1 + 3\frac{\delta l}{l_0}\right).$$

Dividing by eqn. (8.11) gives

$$\frac{\delta V}{V_0} = 3\frac{\delta l}{l_0}. \tag{8.12}$$

Using eqns. (8.9) and (8.10),

$$\delta V = \frac{\partial V}{\partial T} \delta T = V_0 \beta \, \delta T,$$

$$\delta l = \frac{\partial l}{\partial T} \delta T = l_0 \alpha \, \delta T,$$

and substituting into eqn. (8.12), gives

$$\beta = 3\alpha.$$

This is a useful relation because it provides a means of calculating β from values of α, the latter being much the easier of the two to measure.

The coefficient of linear thermal expansion α as defined in eqn. (8.10) varies only slightly with temperature, and for many real processes it can be regarded as being constant in value.[†] It can have positive or negative values.

When α can be taken as a constant, then for a constant stress process eqn. (8.10) can be integrated to give

$$\frac{l - l_0}{l_0} = \alpha T, \tag{8.13}$$

where here l_0 is the fictitious length corresponding to zero temperature.

[†] A nickel–steel alloy has been found to have a very small variation of α with temperature making it a suitable material for the balance-wheels of watches.

8.6 Equations of state

Returning again to the case of a rod under tension, discussion is now limited by the following conditions:

(a) That the rod is under a simple, uniform, tensile stress distribution. Torsion and bending stresses, for example, are absent.
(b) That no body forces are acting on the rod.
(c) That no hydrostatic pressure is acting on the rod.
(d) That all parts of the rod are at zero velocity.
(e) That the rod is homogeneous and invariant in mixture, and in crystal, molecular and atomic structure.
(f) That no chemical or atomic reactions are taking place.
(g) That no electrical current is passing through the rod.

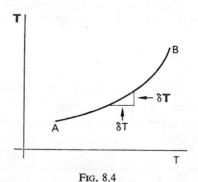

Fig. 8.4

In general, a work and heat process upon the rod will have associated changes of **T** and T during the process. No matter what the form of the process it can be split up into an infinite number of infinitesimal processes, infinitesimal $d\mathbf{T}$ and dT processes occurring alternately as illustrated in Fig. 8.4.

THE FIRST LAW OF THERMODYNAMICS TO SOLIDS 111

In this restricted case, the length of the rod is a function of only **T** and T provided that these quantities are uniform throughout the rod. This is because no other external influences are available except load and heat. A real process taking place at a finite rate will approximate more and more to these requirements of uniform **T** and T as the process becomes slower and slower. Assuming in the first instance that this requirement of slowness is met, then if the function $l = l(T, \mathbf{T})$ is differentiable[†] an incremental change in length, dl, for the two infinitesimal **T** and T processes can be expressed by

$$dl = \left(\frac{\partial l}{\partial T}\right)_{\mathbf{T}} dT + \left(\frac{\partial l}{\partial \mathbf{T}}\right)_{T} d\mathbf{T}.$$

Substituting from eqns. (8.1) and (8.10), this becomes

$$dl = \alpha l_0 \, dT + \frac{l_0}{YA_0} \, d\mathbf{T}. \tag{8.14}$$

Integrating this expression and dividing by l_0 gives

$$\frac{l - l_0}{l_0} = \alpha T + \frac{\mathbf{T}}{YA_0}, \tag{8.15}$$

where now l_0 is the fictitious length when $T = \mathbf{T} = 0$.[‡]

The rod can be thought of as being composed of an infinitesimal number of infinitesimally small rods. As each of these rods shrinks to zero size then so does the distribution of **T** and T within each element approach uniformity even though their gradients remain finite. Thus, in this limit, there is a single value of **T** and of T associated with each element. If α and Y are unaffected by the speed of the process, eqn. (8.15) will then apply individually to each element and for an overall change

[†] See further discussion in § 10.3.

[‡] $(l-l_0)/l_0$ is not the strain because it results from a temperature change as well as from a stress.

in **T** and T will give the corresponding size change of each element. All these elemental size changes are additive and will thus give the total size change of the rod. If, as illustrated in Fig. 8.4, a process takes place at a finite rate between states A and B which are homogeneous ones, that is they are states for which **T** and T are uniform throughout the rod, then eqn. (8.15) will still represent the change of l resulting from the process.

Equation (8.15) thus applies to two cases; first, it applies to any infinitesimal element of the rod at any instant during a process which is proceeding at a finite rate; second, it applies to the change between any two homogeneous states resulting from a process that occurs at a finite rate.

Equation (8.15) gives a unique relation between the properties l/l_0, T, and **T** in equilibrium states or associated with infinitesimal elements. It is called an equation of state. Mathematically, l/l_0, T, and **T** are variables any two of which can be regarded as the independent ones.

Equation (8.15) applies to a material such as steel for which

$$\alpha = 11 \cdot 0 \times 10^{-6} \, °K^{-1}$$

and $$Y = 3 \times 10^7 \, lb \, in^{-2}.$$

Equation (8.15), being linear, changes can be written as

$$\frac{\Delta l}{l_0} = 11 \cdot 0 \times 10^{-6} \, \Delta T + 3 \cdot 3 \times 10^{-8} \frac{\Delta \mathbf{T}}{A_0}.$$

In many engineering applications the stress $\mathbf{T}/A_0$ in lb in^{-2} is much greater than changes in the temperature in °K, and so the second term is the dominant one. However, a rod can be constrained to effectively constant length by a surrounding structure, and then

$$\Delta \left(\frac{\mathbf{T}}{A_0} \right) = -3 \cdot 3 \times 10^2 \, \Delta T.$$

Thus a temperature increase results in a compressive stress.

The equation of state for a strip of rubber is different in form from the one just derived for an elastic metal. First, at constant temperature the relation between the stress $\mathbf{T}/A_0$ and l is not linear as in eqn. (8.2) but, for extensions up to values of l/l_0 of about 2·0, is of the form

$$\frac{\mathbf{T}}{A_0} \propto \left(\frac{l-l_0}{l_0}\right)\left[1-C\left(\frac{l-l_0}{l_0}\right)\right],$$

where C is a positive constant of order $\frac{1}{2}$, and where l_0 is the length at zero stress. Second, the relation between $\mathbf{T}$ and T is linear so that[2],

$$\frac{\mathbf{T}}{A_0} \propto T.$$

The relation that satisfies these two requirements is

$$\frac{\mathbf{T}}{A_0} = kT\left(\frac{l-l_0}{l_0}\right)\left[1-C\left(\frac{l-l_0}{l_0}\right)\right]. \qquad (8.16)$$

The length at zero stress is a function of T in the form of eqn. (8.10). The value of α for rubber is about $2 \cdot 2 \times 10^{-4}$ °K^{-1}. Differentiation of eqn. (8.16) and substitution into eqn. (8.1) results in

$$Y = kT\left[1-2C\left(\frac{l-l_0}{l_0}\right)\right].$$

The value of Young's modulus at zero extension Y_0 is thus

$$Y_0 = kT.$$

A typical value of Y_0 at 15° C is about 10^6 N m^{-2}. With rubber, changes in length due to application of stress are usually very much greater than changes in length, at zero stress, due to temperature changes. Thus l_0 can be regarded as a constant in eqn. (8.16).[†]

[†] It is only because of the introduction of this approximation that use of eqns. (8.10) and (8.16) leads to a zero value for α when $T = 0$ and $l = l_0$. This happens with a metal.

8.7 The two-property substance

It was pointed out in the last section that there are only two independent properties in the equations of state that were derived. In these cases the state of the material is fixed precisely by fixing the value of two properties; such a material is here called a two-property substance.

The state that is determined is defined only in terms of those properties that are relevant to the processes that have been under discussion: for instance the properties discussed would not define a state that could be associated with a process involving the passage of an electric current because the electrical conductivity could be altered by small traces of impurities which would have an unmeasurable effect upon the equation of state relation between $\mathbf{T}, T$, and l: similarly, traces of impurities could markedly affect processes that involve chemical reactions.

The choice of properties as independent variables is not arbitrary. An example is illustrated by the relation between l and the volume V given in eqn. (8.8) which means that these properties are not independent variables. V and l are independent properties when a significant hydrostatic pressure is acting as well as $\mathbf{T}$, but then, for such a process, the material is not a two-property substance.

The two-property substance has been called a pure substance though for reasons just given this usage can be a misnomer.

8.8 The internal energy

The internal energy, being a property, is fixed by a definition of the state. For a two-property substance it is a function of two properties. For this special case the internal energy is given the symbol U. Remembering the discussion of § 5.9, there is value

THE FIRST LAW OF THERMODYNAMICS TO SOLIDS 115

in defining an internal energy per mass unit u by

$$u \equiv \frac{U}{m}. \qquad (8.17)$$

In the present case, two properties that define u are l and T. Then, if discontinuities are excluded so that u is a perfect differential, infinitesimal changes in u can be expressed by

$$du = \left(\frac{\partial u}{\partial T}\right)_l dT + \left(\frac{\partial u}{\partial l}\right)_T dl. \qquad (8.18)$$

The first differential, which is the rate of change of u with temperature at constant length, is given the symbol C_u, thus

$$C_u \equiv \left(\frac{\partial u}{\partial T}\right)_l.$$

The symbol C_V has been used for this quantity by engineers who call it the specific heat, following physicists who have defined C_V as (dQ/dT) for a process in which l is held constant, and who refer to it also as the heat capacity. As an object of the present section is to show that u cannot be associated in general with heat, then C_u will be referred to here as the internal energy coefficient.

For this two-property substance a plot of u against T in the form of lines of constant l will give single-valued curves, such as is sketched in Fig. 8.5. If these curves are differentiable at every point, then C_u is a single-valued function of l and T, and is thus a property.

For a system,

$$mC_u = m\frac{\partial u}{\partial T} = \frac{\partial}{\partial T}(mu) = \frac{\partial U}{\partial T}.$$

If the system is composed of two portions, its mass and internal energy may be denoted by m_1, m_2 and U_1, U_2. As the internal energy is an additive property, then

$$U = U_1 + U_2.$$

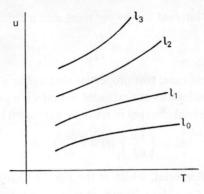

Fig. 8.5

Differentiating,

$$\frac{\partial U}{\partial T} = \frac{\partial U_1}{\partial T} + \frac{\partial U_2}{\partial T}$$

or

$$\frac{\partial}{\partial T}(mu) = \frac{\partial}{\partial t}(m_1 u_1) + \frac{\partial}{\partial t}(m_2 u_2)$$

and so

$$m\frac{\partial u}{\partial t} = m_1 \frac{\partial u_1}{\partial t} + m_2 \frac{\partial u_2}{\partial t}.$$

Therefore

$$mC_u = m_1 C_{u_1} + m_2 C_{u_2},$$

and the quantity (mC_u) is an additive property.

Experimentally C_u is found to vary with temperature. For copper it tends to zero at $0°$ K and is about 355 J kg^{-1} $°$K^{-1} at $273°$ K. But with an increase of temperature to $373°$ K it increases by only another $3\frac{1}{2}$ per cent. The corresponding change for iron is about 9 per cent. The effect of hydrostatic pressure upon C_u is very small for the order of stress commonly occurring in the structural use of solids. Thus for many applications C_u can be regarded as a constant.

The second differential in eqn. (8.18) is expressed by the appropriate version of eqn. (7.12)[3] as

$$\left(\frac{\partial U}{\partial l}\right)_T = -T\left(\frac{\partial \mathbf{T}}{\partial T}\right)_l + \mathbf{T}. \qquad (8.19)$$

8.9 Internal energy of an elastic metal

Where l is constant, $dl = 0$, and so eqn. (8.14) for an elastic metal gives

$$\left(\frac{\partial \mathbf{T}}{\partial T}\right)_l = -\alpha Y A_0.$$

Substitution of this into eqn. (8.19) and noting eqn. (8.15), leads to

$$\left(\frac{\partial U}{\partial l}\right)_T = \alpha Y A_0 T + \mathbf{T}$$
$$= Y A_0 \left(\frac{l-l_0}{l_0}\right).$$

Thus noting that $U = \varrho_0 A_0 l_0 u$, eqn. (8.18) becomes

$$du = C_u \, dT + \frac{Y}{\varrho_0 l_0}\left(\frac{l-l_0}{l_0}\right) dl.$$

If C_u and Y are regarded as constants, this relation can be integrated, giving

$$u = C_u T + \frac{1}{2}\frac{Y}{\varrho_0}\left(\frac{l-l_0}{l}\right)^2 + \text{constant}.$$

Putting $u = 0$, where $T = 0$ and $l = l_0$, then

$$u = C_u T + \frac{1}{2}\frac{Y}{\varrho_0}\left(\frac{l-l_0}{l_0}\right)^2. \qquad (8.20)$$

The internal energy is a function of both T and l. This is why the internal energy is not called heat, for the latter is by its present definition and common usage associated directly with temperatures, whereas the above relation shows that u can change as l changes even though T remains constant.

If the internal energy is referred to an infinitesimal element, then it can be expressed by substituting from eqn. (8.15) into eqn. (8.20), giving

$$u = C_u T + \frac{1}{2} \frac{Y}{\varrho_0} \left[\alpha T + \frac{\mathbf{T}}{YA_0} \right]^2,$$

in which all the quantities can be associated with a point.

8.10 Internal energy of rubber

Proceeding as before, differentiation of the equation of state for rubber, eqn. (8.16), and putting $dl = 0$, gives

$$\frac{1}{A_0} d\mathbf{T} = k \left(\frac{l-l_0}{l_0} \right) \left[1 - C \left(\frac{l-l_0}{l_0} \right) \right] dT,$$

and thus, by making use of eqn. (8.16),

$$\left(\frac{\partial \mathbf{T}}{\partial T} \right)_l = \frac{\mathbf{T}}{T}.$$

Putting this value into eqn. (8.19) results in

$$m \left(\frac{\partial u}{\partial l} \right)_T = \left(\frac{\partial U}{\partial l} \right)_T = -T \frac{\mathbf{T}}{T} + \mathbf{T} = 0,$$

and so eqn. (8.18) becomes

$$du = C_u \, dT. \tag{8.21}$$

Assuming, as before, that C_u is constant, then integration of this relation subject to the condition that $u = 0$, when $t = 0$, gives

$$u = C_u T.$$

Unlike metal, rubber—to the high degree of approximation accepted here—has an internal energy that is a function of temperature only.

8.11 Application of the first law of thermodynamics to the stretching of steel

Substitution of the equation of state for a steel rod, eqn. (8.14), into eqn. (8.3) gives the following expression for the work done in any type of process as

$$W_T = \int_{l_0}^{l} \mathbf{T} \left(\alpha l_0 \, dT + \frac{l_0}{YA_0} \, d\mathbf{T} \right),$$

which for α and Y constant becomes

$$W_T = \alpha l_0 \int_{l_0}^{l} \mathbf{T} \, dT + \frac{l_0}{2YA_0} \mathbf{T}^2. \tag{8.22}$$

This relation is only applicable to processes that are slow enough to maintain the homogeneity implicit in eqn. (8.14). The remaining integral cannot be evaluated until $\mathbf{T}$ is known as a function of T: this again stresses the point that the work done is dependent upon the path followed by the process. However, when a piece of steel is stretched adiabatically, the resulting change of temperature is small. This will now be illustrated. It has been shown by a numerical example in § 8.6 that in this case the temperature term in eqn. (8.15) is negligible, and so, when it is substituted into eqn. (8.20),

$$u = C_u T + \frac{1}{2} \frac{\mathbf{T}^2}{\varrho_0 Y A_0^2}. \tag{8.23}$$

If during such a process the heat applied by the rod is Q_{by}, then in a change in state from state A to state B the first law of thermodynamics, eqn. (6.3), gives

$$\int_A^B (W_{to} - Q_{by}) = m(u_B - u_A).$$

Substitution from eqns. (8.22) and (8.23) into this relation then gives

$$-\Big|_A^B Q_{\text{by}} + \alpha l_0 \int_A^B \mathbf{T}\, dT + \frac{l_0 \mathbf{T}^2}{2YA_0}$$
$$= \varrho_0 A_0 l_0 \left[C_u(T_B - T_A) + \frac{1}{2}\frac{\mathbf{T}^2}{\varrho_0 Y A_0^2} \right].$$

Now

$$\varrho_0 A_0 l_0 C_u(T_B - T_A) = \varrho_0 A_0 l_0 \int_A^B C_u\, dT,$$

and so

$$-\Big|_A^B Q_{\text{by}} = \varrho_0 A_0 l_0 \int_A^B C_u\, dT - \alpha l_0 \int_A^B \mathbf{T}\, dT$$
$$= A_0 l_0 \int_A^B \left(\varrho_0 C_u - \alpha \frac{\mathbf{T}}{A_0} \right) dT.$$

For an adiabatic process for which $\Big|_A^B Q_{\text{by}} = 0$ then

$$\int_A^B \left(\varrho_0 C_u - \alpha \frac{\mathbf{T}}{A_0} \right) dT = 0.$$

As the integrand is a function of $\mathbf{T}$ which varies during this process, then the integral can only be zero when there is no temperature change. Therefore $T_A = T_B$.

A more precise calculation for a tensile stress of $1 \cdot 5 \times 10^4$ lb in^{-2} acting on a mild steel rod gives a fall in temperature of $0 \cdot 1°$K.[†] Details of the more accurate method are given in ref. 3 and experimental confirmation of this small effect is reported in ref. 4.

[†] A compressive stress of the same value would give a rise in temperature of the same amount.

THE FIRST LAW OF THERMODYNAMICS TO SOLIDS   121

The second internal energy term in eqn. (8.23) which, for the process discussed, is seen to be equal to the work per mass unit done by the tensile load, is called the strain internal energy or just the strain energy. This distinguishes it from the first term which is temperature dependent and which might be called the thermal internal energy.

8.12 Application of the first law of thermodynamics to the stretching of rubber

For an infinitesimal part of an adiabatic stretching process on rubber the work done can be written $\mathbf{T}\,dl$ and the internal energy change can, from eqn. (8.21), be written $mC_u\,dT$. Application of the first law of thermodynamics, eqn. (6.3), and putting $Q = 0$, then gives

$$\mathbf{T}\,dl = mC_u\,dT.$$

Eliminating $\mathbf{T}$ by use of eqn. (8.16) gives

$$kA_0\left(\frac{l-l_0}{l_0}\right)\left[1 - C\left(\frac{l-l_0}{l_0}\right)\right]dl = mC_u\frac{dT}{T}.$$

By assuming that C_u is a constant, this can be integrated to give, for a suitably slow process between states A and B,

$$\frac{kA_0l_0}{2}\left(\frac{l-l_0}{l_0}\right)^2\left[1 - \frac{2}{3}C\left(\frac{l-l_0}{l_0}\right)\right] = mC_u\log\frac{T_B}{T_A}.$$

Therefore

$$\frac{k}{2\varrho_0}\left(\frac{l-l_0}{l_0}\right)^2\left[1 - \frac{2}{3}C\left(\frac{l-l_0}{l_0}\right)\right] = C_u\log\frac{T_B}{T_A}. \quad (8.24)$$

This equation provides a relation between the length change during an adiabatic process and the corresponding temperature change. Temperature changes are found to be small, and so the right-hand side of this equation can be written $C_u\,[(T_B/T_A) - 1]$.

Results of experimental measurements on the change in temperature made by Joule[5] and by James and Guth[6] are shown plotted in Fig. 8.6. The observed small drop in temperature at low extensions can be explained by an analysis that accounts for the approximation made here.[6]

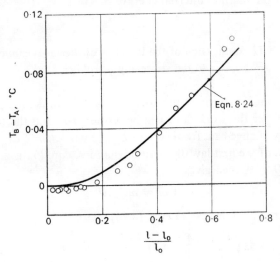

Fig. 8.6

8.13 The rate equation

In the preceding analysis, in which the first law of thermodynamics has been applied to work and heat processes on solids, these processes have been assumed to be very slow ones so that in the evaluation of W and U the temperature and the stress could be assumed to be uniform throughout the solid at all instants during the process. This speed limitation is not necessary as will now be shown.

The heat alone process has already been considered in

THE FIRST LAW OF THERMODYNAMICS TO SOLIDS 123

§ 7.5. In the attendant discussion of § 7.2 it was shown how the infinitesimal element satisfies the requirements of the first law of thermodynamics that the temperature be uniform throughout the system. The work alone process has also been considered in § 8.2 and, again, the infinitesimal element satisfies the first law requirement that the stress be uniform and that there be no relative motion.

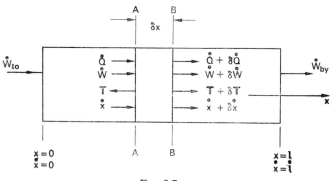

Fig. 8.7

Figure 8.7 is a sketch of such an element of a rod which, at an instant in time, is undergoing a process for which the following details are applicable and for which the discussion is a direct extension of that in § 7.5. Across section A–A heat is being applied to the element at a rate $\dot{Q}$; across B–B heat is being applied by the element at a rate $\dot{Q}+\delta\dot{Q}$. Similar work terms are $\dot{W}$ and $\dot{W}+\delta\dot{W}$. The tensile force acting across section A–A is $\mathbf{T}$ and that across B–B is $\mathbf{T}+\delta\mathbf{T}$. Whilst the rod is being stretched, the element will be moving in the x-direction as sketched; the corresponding velocity of section A–A will be $\dot{x}$ and that of section B–B will be $\dot{x}+\delta\dot{x}$.

The rate at which work is being done to the element at section A–A is given by

$$\dot{W} = -\mathbf{T}\dot{x},$$

the negative sign occurring because **T** and $\dot{x}$ are opposite in direction as sketched in Fig. 8.7. Similarly, at section B–B, the work done by the element is $\dot{W}+\delta\dot{W}$, so that the work done to the element is

$$-(\dot{W}+\delta\dot{W}) = (\mathbf{T}+\delta\mathbf{T})(\dot{x}+\delta\dot{x}),$$

and so the rate at which work is being done to the element is

$$\dot{W}-(\dot{W}+\delta\dot{W}) = -\mathbf{T}\dot{x}+(\mathbf{T}+\delta\mathbf{T})(\dot{x}+\delta\dot{x})$$

or
$$-\delta\dot{W} = \dot{x}\,\delta\mathbf{T}+\mathbf{T}\,\delta\dot{x}, \qquad (8.25)$$

where the second-order term is neglected. The two terms on the right-hand side of this equation have the following physical significance. The first one is the resultant force on the element times the velocity of the element, and so is the rate of doing work in a mechanical process which results in a change in kinetic energy. As $\delta\dot{x}$ is the difference in velocity between sections A–A and B–B, it is the rate at which the element is expanding, and so the second term is the rate at which work is being done in a process that is governed by the first law of thermodynamics.

Supposing that the element goes through a series of processes which, as just described, are alternately mechanical with motion of its centre of mass and thermodynamic without such motion, then, as these processes become smaller in effect and larger in number, in the limit the result is a combination of thermodynamic and mechanical processes proceeding smoothly and simultaneously.

Calling the mass of the element δm, then, for each infinitesimal thermodynamic process taking a time δt, the first law of thermodynamics gives

$$\mathbf{T}\,\delta\dot{x}\,\delta t - \delta\dot{Q}\,\delta t = \frac{D}{Dt}(u\,\delta m)\,\delta t,$$

THE FIRST LAW OF THERMODYNAMICS TO SOLIDS 125

and so
$$\mathbf{T}\,\delta\dot{x} - \delta\dot{Q} = \frac{D}{Dt}(u\,\delta m), \tag{8.26}$$

where the total differential is taken because we are concerned with one particular element and not one particular fixed x position.[†]

For each mechanical process the rate of doing work is, by eqn. (1.17), equal to the rate of change of kinetic energy, and so
$$\dot{x}\,\delta\mathbf{T} = \frac{D}{Dt}\left(\frac{1}{2}\dot{x}^2\,\delta m\right).$$

Adding this relation to eqn. (8.26) gives, for the two processes,
$$\dot{x}\,\delta\mathbf{T} + \mathbf{T}\,\delta\dot{x} - \delta\dot{Q} = \frac{D}{Dt}\left[u\,\delta m + \frac{1}{2}\dot{x}^2\,\delta m\right]$$

and, substituting from eqn. (8.25),
$$-\delta\dot{W} - \delta\dot{Q} = \frac{D}{Dt}\left[\left(u + \frac{1}{2}\dot{x}^2\right)\delta m\right].$$

This applies to every element along the rod, and summing, as in § 7.5, for all elements of a rod of length l, gives
$$\int_0^l -d\dot{W} + \int_0^l -d\dot{Q} = \frac{D}{Dt}\int_0^l \left(u + \frac{1}{2}\dot{x}^2\right)dm. \tag{8.27}$$

As in the discussion of § 7.5, the first summation on the left-hand side becomes, as illustrated by Fig. 8.7,
$$\int_0^l -d\dot{W} = \dot{W}_{x=0} - \dot{W}_{x=l}$$
$$= \dot{W}_{\text{to}},$$

where $\dot{W}_{\text{to}}$ signifies the total rate at which work is being done

[†] This point is amplified later in § 11.8.

126   THERMOMECHANICS

to the complete rod. Similarly, the second term becomes $\dot{Q}_{\text{to}}$, the total rate at which heat is being applied to the rod. Thus eqn. (8.27) becomes

$$\dot{W}_{\text{to}} + \dot{Q}_{\text{to}} = \frac{DE_T}{Dt}$$

or
$$\dot{Q}_{\text{to}} - \dot{W}_{\text{by}} = \frac{DE_T}{Dt}, \qquad (8.28)$$

where E_T, the total energy, is defined by[†]

$$E_T \equiv \int_0^l \left(u + \frac{1}{2} \dot{x}^2 \right) dm. \qquad (8.29)$$

Extending the discussion of § 7.5, eqn. (8.28) can be integrated with respect to time between any two instants during the process to give

$$\left|_{t_1}^{t_2} Q_{\text{to}} - \right|_{t_1}^{t_2} W_{\text{by}} = E_{T,\,t_2} - E_{T,\,t_1}.$$

If, in addition, these two states are ones of stationary equilibrium, A and B, say, then eqn. (6.9) is recovered as

$$\left|_A^B (Q_{\text{to}} - W_{\text{by}}) = U_B - U_A.\right.$$

Returning now to eqn. (8.25), we have,

$$-\delta \dot{W} = \delta(\mathbf{T}\dot{x}),$$

and with $\dot{x} = 0$ at $x = 0$, then

$$\dot{W}_{\text{to}} = \int_0^l d(\mathbf{T}\dot{x})$$
$$= \mathbf{T}_l \dot{l},$$

where $\mathbf{T}_l$ is the value at $x = l$, where the velocity is $\dot{l}$.

[†] E_T is not necessarily the same as E. The first law of thermodynamics has here been based upon equilibrium states in which all properties of this simple system, including e, have a single value throughout the system; here u and $\dot{x}$ can vary.

Then

$$W = \int \dot{W}\,dt = \int \mathbf{T}_l \dot{l}\,dt.$$

When $\mathbf{T}_l$ is a single-valued function of l then this can be written

$$W = \int \mathbf{T}\,dl.$$

The first of the latter two equations is consistent with the formulation of eqn. (5.12) and the latter provides the necessary extension of eqn. (8.3) to account for the presence both of motion and the lack of uniformity of $\mathbf{T}$ along the rod.

References

1. WATSON, S. J., Creep and relaxation with thermal cycling, ch. 13 of *Thermal Stress* (Eds. P. P. Benham and R. Hoyle), Pitman London, 1964.
2. GEE, G., The interaction between rubber and liquids: IX, The elastic behaviour of dry and swollen rubbers, *Trans. Faraday Soc.* **42** (8), 589 (Aug. 1946).
3. ZEMANSKY, M. W., *Heat and Thermodynamics*, 4th edn., McGraw-Hill, New York, 1957, p. 248.
4. JOULE, J. P., On some thermodynamic properties of solids, *Phil. Trans. Roy. Soc. London*, **149** (1859), 100.
5. *Ibid.*, p. 105.
6. JAMES, H. M. and GUTH, E., Theory of elastic properties of rubber, *J. Chem. Phys.* **11** (10), 475 (Oct. 1943).

CHAPTER 9

THE STATE OF MOTIONLESS FLUIDS

9.1 The equation of state of liquids

Remembering the discussion of § 5.5, it is implied in the present section that the effects of body forces upon the pressure distribution in a quantity of liquid can be accounted for, so that their removal results in the pressure being a homogeneous property. Later it will be shown that a further requirement for this to be so is that there is no relative motion within the liquid.

In discussing the equation of state of an elastic solid in § 8.6, the presence of various phenomena was excluded. The phenomenon of surface tension is now included in this list. With this addition the equation of state of a liquid is similar to that of an elastic solid and the density ϱ is a function of only the pressure and the temperature, or

$$\varrho = \varrho(p, T). \tag{9.1}$$

This means that it is a two-property substance.

Usually any one of the three variables in this equation is a single-valued function of the other two. One notable exception to this occurs with water which, at constant pressure, has a maximum density at about 4° C as illustrated in Fig. 9.1. Specifying the pressure and the density could give two solutions for the temperature as indicated by points A and B in this figure.

THE STATE OF MOTIONLESS FLUIDS

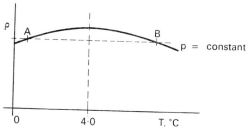

Fig. 9.1

With eqn. (9.1) the compressibility of a liquid C can be defined as

$$C \equiv \frac{1}{\varrho_0}\left(\frac{\partial \varrho}{\partial p}\right)_T, \qquad (9.2)$$

where ϱ is the density, the suffix 0 indicating a reference value, and the differential coefficient being formed at constant temperature.[†] Over moderate ranges of pressure and temperature C is closely constant; for water it is more sensitive to temperature changes than to pressure changes and has a minimum value at about 50° C.

Additionally, a coefficient of thermal expansion β can be defined by the relation[‡]

$$\beta \equiv -\frac{1}{\varrho_0}\left(\frac{\partial \varrho}{\partial T}\right)_p. \qquad (9.3)$$

Noting eqn. (9.1) a small change in density $d\varrho$ can be written

$$d\varrho = \left(\frac{\partial \varrho}{\partial p}\right)_T dp + \left(\frac{\partial \varrho}{\partial T}\right)_p dT. \qquad (9.4)$$

[†] If the liquid was not a two-property substance, then for every additional property added to the independent variables of eqn. (9.1) a further suffix would have to be added to the T-suffix of this differential.

[‡] This is similar to eqn. (8.9) bearing in mind that $\varrho = m/V$ and noting that $\mathbf{T}$ is now replaced by p.

Substituting from eqns. (9.2) and (9.3) this can be rewritten

$$d\varrho = \varrho_0 C\, dp - \varrho_0 \beta\, dT$$

or
$$d\left(\frac{\varrho}{\varrho_0}\right) = C\, dp - \beta\, dT. \qquad (9.5)$$

In integrating this expression it is usually sufficiently accurate

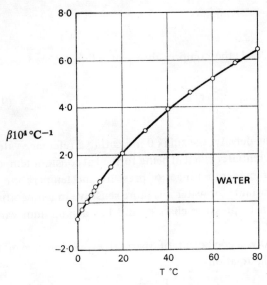

Fig. 9.2

to regard C as a constant. The value of β for water is affected by only extremes of pressure variation, but it is quite sensitive to temperature changes. This is illustrated by a plot, shown in Fig. 9.2, of experimental values.[1,2] So β can only be regarded as a constant if the change in temperature is quite small. When this is so, an integration results in

$$\frac{\varrho - \varrho_0}{\varrho_0} = Cp - \beta T, \qquad (9.6)$$

where now ϱ_0 is an imaginary density at zero pressure and temperature.† If this equation is used to compute changes in density, then substitution of a value of β corresponding to the mid-point of the corresponding temperature range, gives a good approximation to an exact integration of eqn. (9.5). This is illustrated in Fig. 9.3 again for water, where the product of the mean value of β and the temperature is compared with $\int_0^T \beta \, dT$.‡

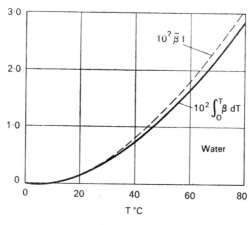

Fig. 9.3

Insertion of numerical values for water into eqn. (9.6) enables a comparison to be made of the relative values of its terms. Typical values of C and β for water are

$$C = 2 \cdot 3 \times 10^{-3} \text{ ft}^2 \text{ lb}^{-1}$$

and

$$\beta = 3 \times 10^{-4} \text{ °K}^{-1},$$

† The reason for this density being an imaginary one will become apparent in § 10.4.

‡ The integration was performed by numerical analysis using Simpson's rule.[3]

giving

$$\frac{\varrho - \varrho_0}{\varrho_0} = 2 \cdot 3 \times 10^{-8} p - 3 \times 10^{-4} T.$$

This shows that a pressure change must be very large to have any significant effect upon the density; a 1 per cent change in density requires a pressure change of $0 \cdot 4 \times 10^6$ lb ft^{-2} which is 190 times atmospheric pressure. In contrast, a similar effect is obtained by a moderate temperature change of 33° C.

9.2 The equation of state of gases

Under the conditions previously laid down in the discussion of the equation of state of liquids, the results now to be described have been obtained experimentally for gases.

In Fig. 9.4 data for the density ϱ of various gases[4] is plotted against the corresponding value of the molecular molar mass M. These results are for a pressure of 1 atm or $2 \cdot 117 \times 10^3$ lb ft^{-2},

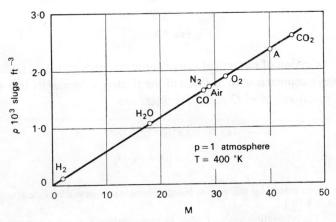

Fig. 9.4

and for a temperature of 400° K.[†] They show that for these experimental conditions

$$\varrho \propto M. \tag{9.7}$$

The density is, by definition, the ratio of a mass to a volume; when the mass is numerically equal to M grams the quantity of the gas is called a mole and then the volume of the gas in cm³ is the volume of 1 mole which is now given the symbol v. Thus the density in gm cm⁻³ is given by

$$\varrho = \frac{M}{v}. \tag{9.8}$$

In Fig. 9.5 the density of hydrogen (H_2) is plotted against the pressure for a temperature of 600° K. It is seen that, over the range shown,[‡]

$$\varrho \propto p. \tag{9.9}$$

Again, in Fig. 9.6 the density of hydrogen is plotted against the temperature for a pressure of 0·01 atm, and this shows that

$$\varrho \propto 1/T. \tag{9.10}$$

Combining eqns. (9.7), (9.9), and (9.10), gives

$$\varrho \propto \frac{pM}{T} \quad \text{or} \quad \frac{pM}{\varrho T} = \text{constant}.$$

The constant, which is independent of the nature of the gas, is called the universal gas constant, and given the symbol R. Thus

$$\frac{pM}{\varrho T} \equiv R \tag{9.11}$$

or, alternatively, from eqn. (9.8),

$$\frac{pv}{T} = R. \tag{9.12}$$

[†] The atmosphere is a unit of pressure defined later in § 9.5.
[‡] On this log–log plot, straight lines at 45° give this result.

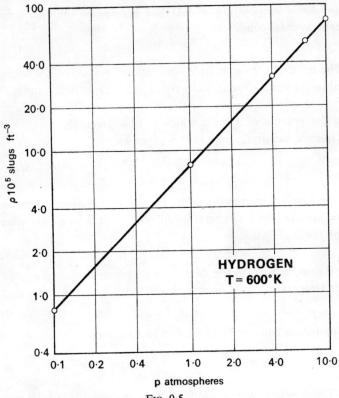

Fig. 9.5

The units of p are force per unit area or in the MKS system, Nm^{-2}; the units of v are volume per mole or $m^3\ mole^{-1}$; and the units of T are °K. Thus the units of R are $Nm^{-2}\ m^3\ mole^{-1}\ °K^{-1}$ or $Nm\ mole^{-1}\ °K^{-1}$. Now the Newton-metre is called the Joule (J), and so[†]

$$R \triangleq J\ mole^{-1}\ °K^{-1}$$

and the numerical value is[(5)]

$$R = 8.314\ J\ mole^{-1}\ °K^{-1}.$$

[†] The symbol $\triangleq$ means "is dimensionally equal to".

For hydrogen gas (H$_2$) $M = 2 \times 1{\cdot}008 = 2{\cdot}016$, and so it has a mass of $2{\cdot}016$ g mole^{-1} or $2{\cdot}016 \times 10^{-3}$ kg mole^{-1}. Thus for hydrogen

$$R/M = 4{\cdot}13 \times 10^3 \text{ J kg}^{-1} \text{ °K}^{-1}$$
$$\triangleq \text{Nm kg}^{-1} \text{ °K}^{-1}$$
$$\triangleq \text{kg m}^2 \text{ sec}^{-2} \text{ kg}^{-1} \text{ °K}^{-1}$$
$$\triangleq \text{m}^2 \text{ sec}^{-2} \text{ °K}^{-1}.$$

As 1 m equals $1/0{\cdot}9144$ yd; then, in English units,

$$R/M = 4{\cdot}13 \times 10^3 \left(\frac{3}{0{\cdot}9144}\right)^2$$
$$= 4{\cdot}44 \times 10^4 \text{ ft}^2 \text{ sec}^{-2} \text{ °K}^{-1}.$$

Similarly for dry air, which at sea-level has a molecular molar mass of $29{\cdot}0$,

$$R/M = 3{\cdot}08 \times 10^3 \text{ ft}^2 \text{ sec}^{-2} \text{ °K}^{-1}.$$

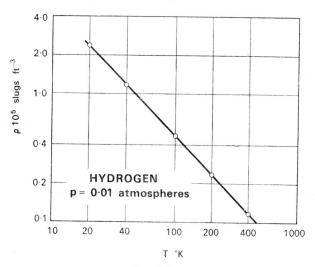

Fig. 9.6

In general, eqn. (9.9) is an approximation to the behaviour of gases. Defining a compressibility factor Z by

$$Z \equiv \frac{pM}{\varrho TR},\qquad(9.13)$$

values of Z for hydrogen are shown plotted in Fig. 9.7 for modest temperatures and over a range of pressure of from 0·01 to 100 atm.[4] Deviation from the law represented by

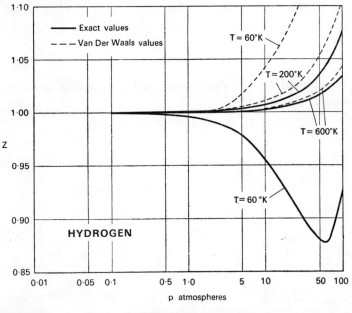

Fig. 9.7

eqn. (9.11) is shown by deviation of the value of Z from unity. Significant deviations are seen to result from an increase of pressure above about atmospheric pressure for the low temperature and for pressure above about 10 atm at the two higher temperatures. The nature of these deviations is also such that the values of Z for $T = 600°$ K lie between those for $60°$ K and $200°$ K.

Thus eqn. (9.11) describes a limiting condition as the pressure tends to zero or, from this equation, for finite temperature the limiting condition is reached as the density tends to zero. For moderate pressures it is possible to write

$$\frac{pM}{\varrho RT} = 1 + a_1\varrho + a_2\varrho^2 + \ldots, \tag{9.14}$$

where a_1, a_2, and so on are functions of temperature.[†] This equation is called the virial equation and the coefficients a_1, a_2, and so on are known as the virial coefficients. As well as equations of the form of eqn. (9.14), many others of different form have been suggested. One of the best of these is that proposed by van der Waals which is[7]

$$\left(p + \frac{a}{v^2}\right)(v - b) = RT,$$

where a and b are constants for any particular gas. Using the values of a and b for hydrogen[8] computation of Z gives the results shown as dotted lines in Fig. 9.7. There is seen to be modest agreement at the two higher temperatures, but a failure of the van der Waals equation for a temperature of $60°$ K. As with all formulae of this type, before using van der Waals' equation the range of its validity for a specified accuracy must be known.

The mass of a single molecule is given by the product of the molecular molar mass M and the unit atomic mass A_m.[‡] The total mass of a volume V of gas containing N molecules is then NA_mM. The density is thus given by

$$\varrho = \frac{NA_mM}{V}. \tag{9.15}$$

[†] It is equally possible to express this as a series in p but it can then happen that the convergence is not so rapid.[6]

[‡] See Chapter 4 for discussion and value of A_m.

However, eqn. (9.7) shows that under specified conditions

$$\frac{\varrho}{M} = \text{constant}.$$

Remembering that A_m is a constant and combining these two relations, gives

$$N \propto \frac{\varrho V}{M} \propto V.$$

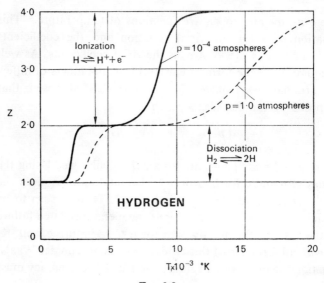

Fig. 9.8

Thus when eqn. (9.7) applies, the number of molecules in a specified volume of gas at a fixed temperature and pressure is the same for all gases. If this result remained true for a gas when its molecules were subdivided, then as the number of particles was increased by this division the volume would correspondingly increase so that the density would decrease. If, in eqn. (9.13), M retained its original value, then after division Z would increase. Figure 9.8, a graph of computed curves for hydrogen based upon experimental data,[9] shows that this happens. As the temperature of the hydrogen gas is raised, the increased energy of the molecular motion results in dissociation of the hydrogen molecules according to the reaction $H_2 \rightleftharpoons 2H$; when dissociation is completed, the number of particles is doubled and so Z attains a value of 2·0. Further increase of temperature results in the hydrogen H, atoms shedding their electrons according to the ionization process $H \rightleftharpoons H^+ + e^-$; again, when ionization is completed, the number of particles is doubled and Z rises to a final value of 4·0. The temperatures at which these effects occur depend upon the pressure. This is illustrated in Fig. 9.8.

A study of Figs. 9.7 and 9.8 and similar graphs for other gases shows that for any particular gas any one of the variables p, ϱ, and T is a function of only the other two. Thus gases are two-property substances.

9.3 The gas thermometer

It has been assumed in the last section that it is possible to measure the temperature T of a gas on the absolute scale. Suppose for a moment that it is not possible. Then measurements of a number of quantities of different gases which are all in thermal equilibrium with each other would show that they all have the same numerical value of pM/ϱ as $p \to 0$. A new level of thermal equilibrium would result in another value of pM/ϱ that is again common to all the gases. The quantity pM/ϱ being composed of properties is thus itself a property. Hence, the quantity pM/ϱ as $p \to 0$ can be used as a numerical measure of temperature.

Referring to eqn. (3.1), a temperature scale can now be defined by

$$T \equiv a + b \left(\frac{pM}{\varrho}\right)_{p \to 0}. \qquad (9.16)$$

Putting $a = 0$, as described in § 4.6, and identifying b with $1/R$ reduces this defining equation to eqn. (9.11). This shows that in effect R is a units conversion factor.

The quantity of gas from which measurements could be taken in this way forms what is known as a gas thermometer. The important feature of the gas thermometer is that it can be shown[10] that the value of the constant b in eqn. (9.16) is directly proportional to the absolute scale of temperature. The gas thermometer is the standard instrument that is used to measure absolute temperatures.

9.4 Partial pressures

Combination of eqns. (9.11) and (9.15) gives

$$p = \frac{NA_mRT}{V}. \qquad (9.17)$$

It has been shown that though a gas can be composed of the different types of particles resulting from dissociation and ionization, there are still the same number of particles N in a specified volume V. This also applies with a mixture of gases.

Suppose two gases, 1 and 2, having numbers of molecules denoted respectively by N_1 and N_2, are mixed together, then

$$N = N_1 + N_2$$

and the above relation becomes

$$p = \frac{(N_1+N_2)A_mRT}{V}. \qquad (9.18)$$

The pressures when each gas is on its own are

$$p_1 = \frac{N_1A_mRT}{V} \quad \text{and} \quad p_2 = \frac{N_2A_mRT}{V}.$$

Substitution of these relations into eqn. (9.18) gives

$$p = p_1 + p_2. \tag{9.19}$$

The quantities p_1 and p_2 are called the partial pressures, and eqn. (9.19) shows that they can be simply added to give the pressure p.

9.5 Pressure distribution in a liquid under gravity

The equation of state for a liquid, eqn. (9.6), can be combined with that for the variation of pressure vertically in a gravitational field, eqn. (5.3), enabling the latter to be integrated. It has already been shown that for the majority of cases eqn. (9.6) can be reduced to the result

$$\varrho = \text{constant}.$$

Thus eqn. (5.3) integrates to

$$p = -\varrho g z + \text{constant} \tag{9.20}$$

and the pressure decreases linearly with vertical increase in height upwards.

This relation enables another unit of pressure, the atmosphere, to be defined as that pressure difference corresponding to a height of mercury of 760 mm. To complete the definition, values of the density of mercury and of g must be added, and it is here that variations in the definition exist making it unsatisfactory. In this book a value of $1 \cdot 013 \times 10^5$ Nm^{-2} for unit atmosphere is used.

An example to which eqn. (9.20) can be applied is illustrated in Fig. 9.9 which is a sketch of a water storage tank supplied by a vertical pipe. The pressures and forces upon the tank are now to be computed for the case shown sketched where the tank has become full and the vertical pipe has also filled up to a height of 30·0 ft. Atmospheric pressure p_a will act upon the

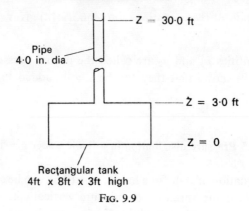

Fig. 9.9

outer surfaces of the tank and pipe and upon the water surface at the upper end of the pipe. Applying eqn. (9.20) to the conditions at the upper water surface that is in contact with the atmosphere gives

$$p_a = -g\varrho 30 + \text{constant},$$

and for water $\varrho = 1\cdot 94$ slugs ft^{-3} so that with $g = 32\cdot 2$ ft sec^{-2},

$$\text{constant} = p_a + 1\cdot 87 \times 10^3 \text{ lb ft}^{-2}.$$

The pressure on the lower surface of the tank is, by eqn. (9.20),

$$p = p_a + 1\cdot 87 \times 10^3.$$

The total force is then

$$(p_a + 1\cdot 87 \times 10^3) 4 \times 8 = (p_a + 1\cdot 87 \times 10^3) 32.$$

However, atmospheric pressure is acting upon the underneath of the lower surface giving a force of $p_a 32$. Thus the net force upon the lower surface is, vertically downward, of amount,

$$(p_a + 1\cdot 87 \times 10^3) 32 - p_a 32 = 5\cdot 97 \times 10^4 \text{ lb}.$$

Similarly, on the upper surface of the tank, the pressure is

$$p = -\varrho g 3 + p_a + 1\cdot 87 \times 10^3$$
$$= p_a + 1\cdot 69 \times 10^3.$$

THE STATE OF MOTIONLESS FLUIDS 143

The area being $32 - \dfrac{\pi}{4}\dfrac{1}{3^2} = 31\cdot 9$, the net force due to this pressure acting upon the inside surface and atmospheric pressure acting upon the outside is

$$[(p_a + 1\cdot 69 \times 10^3) - p_a]\,31\cdot 9 = 5\cdot 39 \times 10^4 \text{ lb.}$$

To compute the force acting upon a side-wall of the tank, the surface is divided into a number of horizontal strips as shown sketched in Fig. 9.10. Each strip, of area $4 \times \delta z$, has

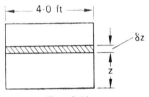

Fig. 9.10

acting upon it a pressure p given by eqn. (9.20) above so that the force acting upon this element is

$$p4\,\delta z = [-\varrho gz + p_a + 1\cdot 87 \times 10^3]\,4\,\delta z$$

The total force is obtained by summing all these forces on the incremental strips to give

$$\text{Total force} = \int_0^{3\cdot 0} [-\varrho gz + p_a + 1\cdot 87 \times 10^3]\,4\,dz$$
$$= 4\left[-\varrho g\,\dfrac{3^2}{2} + 3p_a + 1\cdot 87 \times 10^3 \times 3\right]$$
$$= 21\cdot 3 \times 10^3 - 12p_a.$$

On the outside a force of $12p_a$ is being exerted by the atmosphere and so the net force is $21\cdot 3 \times 10^3$ lb. The forces on the other surfaces can be computed in a similar way.

The total weight of water contained is

$$\varrho g\left[4 \times 8 \times 3 + 27\,\dfrac{\pi}{4}\,\dfrac{1}{3^2}\right] = 6\cdot 14 \times 10^3 \text{ lb,}$$

and this is equal to the difference between the forces acting upon the upper and lower surfaces which is $(5 \cdot 97 - 5 \cdot 39)10^4 = 5 \cdot 8 \times 10^3$ lb.[†]

It is worth noting that the forces on the surfaces are not directly related to the weight of water contained. Removing the small proportion of water contained in the pipe has a very large effect in that it reduces the net load upon the upper surface to zero and reduces that upon the lower surface to one-tenth part of its original value.

9.6 Forces on surfaces immersed in liquids

The preceding numerical example is now extended to consider, in more general terms, the forces acting upon surfaces immersed in liquids.

A plane surface immersed below a liquid surface is shown sketched in Fig. 9.11. As before it is divided into horizontal strips. Putting $z = 0$ at the liquid surface means that the value of z corresponding to the level of the strip will be negative. The surface being inclined at an angle θ to the horizontal, as sketched, then

$$-z = y \sin \theta,$$

and so the pressure acting on the strip will be

$$p = p_a + g\varrho y \sin \theta.$$

[†] It would have been more impressive if the usual practice adopted in textbooks had been employed here. That is, if all calculations had been carried out with great accuracy so that these two results had agreed to about three significant figures. All the present calculations have been performed upon a slide-rule as is almost always done in practice. In the present instance this has the advantage that it brings out a common numerical difficulty arising when an answer is formed by the small difference of two large quantities.

THE STATE OF MOTIONLESS FLUIDS

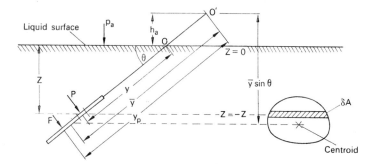

Fig. 9.11

This expression can be written in the form

$$p = g\rho y' \sin \theta,$$

where

$$y' = y + \frac{p_a}{g\rho \sin \theta}.$$

The physical significance of the term $p_a/g\rho \sin \theta$ can be seen as follows. The atmospheric pressure p_a can be thought of as being equivalent to an added height h_a of liquid above the free surface, given by

$$p_a = g\rho h_a$$

and as sketched in Fig. 9.11. This provides an imaginary origin O' the distance $O-O'$ being $h_a/\sin \theta = p_a/g\rho \sin \theta$.

The area of the strip being denoted by δA, the force on it will be

$$(g\rho y' \sin \theta) \delta A.$$

Summing this will give the total force F as

$$F = \int (g\rho y' \sin \theta) \, dA,$$

the integration being taken from the top to the bottom of the surface. This can be rewritten as

$$F = g\rho \sin \theta \int y' \, dA.$$

The integral is the first moment of area about O' and can be written as

$$\int y' \, dA = \bar{y}A, \qquad (9.21)$$

where $\bar{y}$ is the distance from O' to the centroid.[†] Thus the expression for the force becomes

$$F = g\varrho\bar{y} \sin\theta \, A. \qquad (9.22)$$

Now $\bar{y} \sin\theta$ is the depth of the centroid, and thus $g\varrho\bar{y} \sin\theta$ is the pressure at this position. It thus happens that the force is equal to the pressure at the centroid times the area; but the total force does not act at this position as will now be shown.

To find the point of action of F, moments about O' of the force upon each element are summed.

The moment upon one element is

$$p \, \delta A \, y' = g\varrho y' \sin\theta \, y' \, \delta A.$$

The total moment is thus

$$\int g\varrho \sin\theta \, y'^2 \, dA,$$

and this can be equated to Fy_p, where y_p is the distance from O' to the point of action of F. Substitution from eqn. (9.22) then gives

$$y_p = \frac{g\varrho \sin\theta \int y'^2 \, dA}{g\varrho \sin\theta \, \bar{y}A}.$$

The integral is the second moment of area about the point O'. From elementary statics this can be written

$$\int y'^2 \, dA = \bar{y}^2 A + k^2 A,$$

where k is the radius of gyration of the plane surface about the horizontal axis through its centroid. Thus

$$y_p = \frac{\bar{y}^2 A + k^2 A}{\bar{y}A}$$

$$= \bar{y} + \frac{k^2}{\bar{y}}. \qquad (9.23)$$

[†] Or, as it is commonly called, "the centre of gravity", a misnomer in this case.

THE STATE OF MOTIONLESS FLUIDS 147

The centre of pressure, the point at which the force acts, is thus always lower than the centroid by the amount $k^2/\bar{y}$.

It is left as an exercise to show that when the pressure p acts on one side of the plate and on the other the atmospheric pressure p_a is acting, then eqns. (9.22) and (9.23) are valid for the resulting net force and its point of action provided that now $\bar{y}$ and y_p are measured from O.

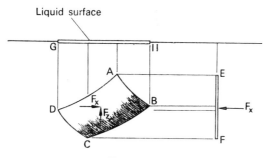

Fig. 9.12

The force on a curved surface, such as that marked *ABCD* in Fig. 9.12, can be obtained by splitting it into horizontal and vertical components, F_x and F_z. An horizontal projection of the perimeter onto a vertical plane *EF* encloses a volume of fluid that is in equilibrium under F_x and the force on the plane surface *EF*. Thus F_x must be equal and opposite to the force on the plane *EF* and also must act along the same line; it can thus be determined using the preceding analysis. A vertical projection of the surface *ABCD* on to the liquid surface encloses a volume of fluid contained between *ABCD* and *GH*. This fluid is in equilibrium under F_z and its own weight. Thus F_z must be equal and opposite to this weight force and must act along the same line.

9.7 Pressure distribution in a gas under gravity

The conditions for which a gas pressure can be regarded as homogeneous were discussed in § 5.5. When height changes are large, such as occurs through the earth's atmosphere, the pressure and also the density vary significantly and their relation must be known before eqn. (5.3) can be integrated.

It will be shown later in § 16.1 that when a gas that obeys eqn.(9.11) and the later relation eqn. (9.31), moves in a process such that there is no heat and no net work done by viscous shear stresses, then the pressure and density are related by

$$p/\varrho^\gamma = k,$$

where γ and k are constants. Substituting this into eqn. (5.3) gives

$$dz = -\frac{k^{\frac{1}{\gamma}}}{g} \frac{dp}{p^{\frac{1}{\gamma}}},$$

which integrates to give

$$z = -\frac{k^{\frac{1}{\gamma}}}{g} \frac{1}{1-(1/\gamma)} p^{1-(\frac{1}{\gamma})} + \text{constant}$$

$$= -\frac{1}{g} \frac{\gamma}{\gamma-1} \frac{p}{\varrho} + \text{constant}.$$

Further substitution from eqn. (9.11) results in

$$z = -\frac{1}{g} \frac{\gamma}{\gamma-1} \frac{R}{M} T + \text{constant}.$$

Thus the temperature decreases linearly with increase in height.

For air,

$$\gamma = 1\cdot 4,$$
$$R/M = 2\cdot 87 \times 10^2 \text{ m}^2 \text{ sec}^{-2} \text{ °K}^{-1}$$

and $g = 9.81$ m sec^{-2}.

Thus $\dfrac{1}{g} \dfrac{\gamma}{\gamma-1} \dfrac{R}{M} = 1.02 \times 10^2$ m °K^{-1},

so that the temperature decreases with height at a rate of $1/1.02 \times 10^2 = 0.98 \times 10^{-2}$ °K m^{-1} or $9.8°$ K/km. In the real atmosphere the temperature does vary linearly with height up to about 11 km over Europe, but it falls at the rate of about $6.5°$ K/km, which is lower than the previously calculated amount. This discrepancy is partly explained as follows and in § 9.11.

This vertical temperature gradient results in heat being applied upwards across any horizontal plane. From eqn. (7.1) the rate of heat per unit area is

$$\eta = -k \frac{\partial T}{\partial z}.$$

For air at 15° C $k = 6.05 \times 10^{-5}$ cal cm^{-1} sec^{-1} °K^{-1}
$= 4.19 \times 6.05 \; 10^{-5}$ J cm^{-1} sec^{-1} °K^{-1}
$= 4.19 \times 6.05 \; 10^{-3}$ J m^{-1} sec^{-1} °K^{-1}
$= 2.54 \times 10^{-2}$ J m^{-1} sec^{-1} °K^{-1},

so that
$\eta = 2.54 \times 10^{-2} \times 0.98 \times 10^{-2} = 2.46 \times 10^{-4}$ J m^{-2} sec^{-1}.

The sun radiates heat to the earth[11] at a rate of about 0.15 W cm^{-2} which is 1.5×10^3 J m^{-2} sec^{-1}. Some of this is absorbed by the atmosphere but there is a large imbalance by a factor of 10^7 at the earth's surface. A result is that a quantity of air there increases in internal energy content with time.

9.8 The internal energy of a liquid

In analogy with eqn. (7.12) it is possible to write for a fluid[12]

$$\left(\frac{\partial U}{\partial V}\right)_T = T \left(\frac{\partial p}{\partial T}\right)_V - p.$$

Remembering that $U = um$ and $\varrho = m/V$ so that $dU = m\,du$ and $dV = -m\,d\varrho/\varrho^2$, this relation can be written

$$-\varrho^2 \left(\frac{\partial u}{\partial \varrho}\right)_T = T\left(\frac{\partial p}{\partial T}\right)_\varrho - p. \qquad (9.24)$$

From the equation of state, eqn. (9.6),

$$\frac{1}{\varrho_0}\,d\varrho = C\,dp - \beta\,dT,$$

so that for constant density

$$\left(\frac{\partial p}{\partial T}\right)_\varrho = \frac{\beta}{C}.$$

Thus, noting eqn. (9.6), eqn.) (9.24) becomes

$$-\varrho^2 \left(\frac{\partial u}{\partial \varrho}\right)_T = \frac{T\beta}{C} - p$$

$$= -\frac{1}{C}\frac{\varrho - \varrho_0}{\varrho_0}.$$

A change in internal energy can, analogously to eqn. (8.18), be written

$$du = \left(\frac{\partial u}{\partial T}\right)_\varrho dT + \left(\frac{\partial u}{\partial \varrho}\right)_T d\varrho. \qquad (9.25)$$

The first differential is given the symbol C_u by the definition

$$C_u \equiv \left(\frac{\partial u}{\partial T}\right)_\varrho, \qquad (9.26)$$

where C_u is here called the internal energy coefficient.[†] The value of C_u for liquids can be regarded as a constant in most calculations. For water at atmospheric pressure and over the

---

[†] Holding the density constant is equivalent to holding the volume constant. Thus $\left(\frac{\partial u}{\partial T}\right)_\varrho = \left(\frac{\partial u}{\partial T}\right)_V \equiv C_V$ and this is the common form of this definition of what is usually called the coefficient of specific heat at constant volume (see § 8.8).

temperature range of 0–100° C it varies plus and minus about 0.5 per cent around a mean value of $4 \cdot 2 \times 10^3$ J kg^{-1} °K^{-1}.

Thus rewriting eqn. (9.25) as

$$du = C_u \, dT + \frac{1}{C} \frac{\varrho - \varrho_0}{\varrho^2 \varrho_0} \, d\varrho$$

it can be integrated to give

$$u = C_u T + \frac{1}{C} \left[\frac{1}{\varrho_0} \log \varrho + \frac{1}{\varrho} \right] + \text{constant}.$$

Putting $u = 0$ when $T = 0$ and $\varrho = \varrho_0$ results in

$$u = C_u T + \frac{1}{C \varrho_0} \left[\log \frac{\varrho}{\varrho_0} + \frac{\varrho_0}{\varrho} - 1 \right].$$

It has already been demonstrated that changes of density in liquids are usually very small, thus

$$\log \frac{\varrho}{\varrho_0} \simeq \frac{\varrho}{\varrho_0} - 1,$$

and substituting this results in

$$u = C_u T + \frac{1}{C \varrho} \left(\frac{\varrho - \varrho_0}{\varrho_0} \right)^2. \tag{9.27}$$

Furthermore it was pointed out in § 9.1 that

$$\frac{\varrho - \varrho_0}{\varrho_0} \simeq -\beta T$$

so that

$$u = C_u T + \frac{\beta^2 T^2}{C \varrho}. \tag{9.28}$$

For water $\beta = 3 \times 10^{-4}$ °K^{-1}, $C = 2 \cdot 3 \times 10^{-8}$ ft^2 lb^{-1}, $\varrho = 1 \cdot 94$ slugs ft^{-3}, and for an atmospheric temperature of 288° K then,

$$\frac{\beta^2 T^2}{C \varrho} = 1 \cdot 7 \times 10^5 \text{ ft}^2 \text{ sec}^{-2}.$$

Also
$$C_u = 4 \cdot 2 \times 10^3 \text{ m}^2 \text{ sec}^{-2} \text{ °K}^{-1}$$
$$= 4 \cdot 2 \times 10^3 \left(\frac{3}{0 \cdot 9144}\right)^2$$
$$= 4 \cdot 5 \times 10^4 \text{ ft}^2 \text{ sec}^{-2} \text{ °K}^{-1}$$

and so
$$C_u T = 1 \cdot 3 \times 10^7 \text{ ft}^2 \text{ sec}^{-2}.$$

Thus, usually, this is the predominant term in eqn. (9.28) and then it is sufficiently accurate to write

$$u = C_u T. \tag{9.29}$$

9.9 The internal energy of a gas

Differentiating the equation of state of a gas, eqn. (9.14), and remembering that its coefficients are functions of only the temperature, results in

$$\left(\frac{\partial p}{\partial T}\right)_\varrho = \frac{\varrho R}{M}[1+(a_1+a_1'T)\varrho+(a_2+a_2'T)\varrho^2+ \ldots],$$

where $a_1' \equiv \dfrac{da_1}{dT}$ and so on. Substituting this and eqn. (9.14) into eqn. (9.24) gives

$$-\varrho^2\left(\frac{\partial u}{\partial \varrho}\right)_T = \frac{\varrho RT}{M}[1+(a_1+a_1'T)\varrho+(a_2+a_2'T)\varrho^2+ \ldots]$$
$$-\frac{\varrho RT}{M}[1+a_1\varrho+a_2\varrho^2+ \ldots].$$

Therefore
$$\left(\frac{\partial u}{\partial \varrho}\right)_T = \frac{RT^2}{\varrho M}[a_1'\varrho+a_2'\varrho^2+ \ldots].$$

The quantity in the brackets on the right-hand side of this equation can be regarded as a differential coefficient with respect to T, the density being held constant; from eqn. (9.14) it can be written

$$\left[\frac{\partial}{\partial T}\left(\frac{pM}{\varrho RT}-1\right)\right]_\varrho$$

so that

$$\left(\frac{\partial u}{\partial \varrho}\right)_T = \frac{RT^2}{\varrho M}\left[\frac{\partial}{\partial T}\left(\frac{pM}{\varrho RT}-1\right)\right]_\varrho.$$

Over the range for which eqn. (9.11) is valid,

$$\frac{pM}{\varrho RT} = 1$$

and thus

$$\left(\frac{\partial u}{\partial \varrho}\right)_T = 0.$$

If the internal energy, which for this two-property substance is a function of any two properties, is regarded as a function of temperature and density, this relation shows that in fact it does not vary with density and so must be a function of only the temperature.

Thus from eqns. (9.25) and (9.26)

$$du = C_u\, dT. \tag{9.30}$$

The internal energy being a function of only the temperature, then C_u must be a function of only the temperature. The degree to which this is so is illustrated in Fig. 9.13, where C_u/R for hydrogen is plotted against the pressure in atmospheres.[4] It is seen that for pressures up to 100 atm, C_u for hydrogen is indeed effectively independent of pressure and so is a function of temperature only. Its variation with temperature is illustrated in Fig. 9·14, where smoothed values from experimental results are plotted up to $600°$ K,[4] and computed values are continued to $1000°$ K.[9] It is seen that C_u is closely a constant over the range of temperature between $300°$ K and

1000° K. At higher temperatures, calculation indicates that C_u has maxima and minima, the curves at different pressures having multiple intersections. Thus because of this latter feature, whilst C_u is a property it cannot be regarded generally as an independent property for uniquely determining the state

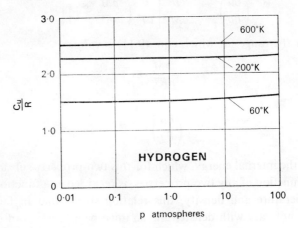

Fig. 9.13

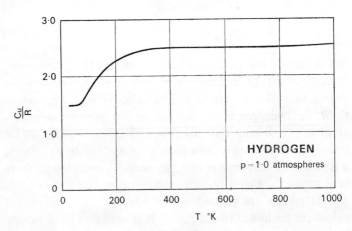

Fig. 9.14

THE STATE OF MOTIONLESS FLUIDS 155

of the gas. The result that, over a range of pressures and temperatures of great practical interest, C_u is closely a constant is a feature common to gases. Using this result, eqn. (9.30) immediately integrates to give

$$u = C_u T + \text{constant}$$

and arbitrarily putting $u = 0$ when $T = 0$ makes this

$$u = C_u T. \tag{9.31}$$

When a gas satisfactorily obeys both eqns. (9.11) and (9.31) it is called an ideal gas.

9.10 Components of a pressure force

Figure 9.15 illustrates a force δF acting upon a small area δA drawn in a fluid; its x-, y-, and z-components are denoted δF_x, δF_y, and δF_z.

If this force arises solely from the existence of a pressure p, then its value is given by

$$\delta F = p\, \delta A$$

and its line of action is perpendicular to δA.

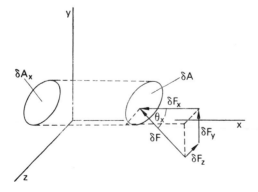

Fig. 9.15

In this figure, δA is shown projected upon a y–z plane to give a surface of area δA_x. The direction cosine being $\cos \theta_x$, as shown sketched, then,

$$\delta A_x = \delta A \cos \theta_x.$$

The x-component of the force δF_x is given by

$$\begin{aligned}\delta F_x &= \delta F \cos \theta_x \\ &= p\, \delta A \cos \theta_x \\ &= p\, \delta A_x.\end{aligned} \qquad (9.32)$$

Similar expressions for the y- and z-components are derived in the same way.

9.11 Pressure forces on fluid volumes

Figure 9.16 is a sketch of a quantity of fluid of volume V. If the pressure varies throughout the fluid, pressure gradients will exist. The rate of change of pressure in the x-direction only, called the x-component of the pressure gradient, is written

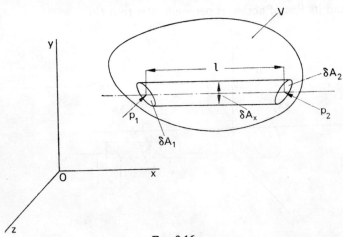

Fig. 9.16

$\partial p/\partial x$. Similarly, in the y- and z-directions the gradients are written $\partial p/\partial y$ and $\partial p/\partial z$ respectively. A small cylindrical element, drawn in the volume and aligned in the x-direction as sketched in this figure, will have pressure forces acting upon each end. Using the notation indicated in Fig. 9.16, the force on the left-hand end is $p_1 \delta A_1$, and from eqn. (9.32) the component of this force in the x-direction is $p_1 \delta A_x$. Similarly, at the other end there is a force in the x-direction of $-p_2 \delta A_x$. The pressures acting upon the side walls of the cylinder have a direction that is perpendicular to Ox, and so the net pressure force in the x-direction is

$$(p_1 - p_2) \delta A_x.$$

The volume of the cylinder δV is given by

$$\delta V = l \, \delta A_x$$

so that the net force can be written

$$\frac{p_1 - p_2}{l} \delta V.$$

There are now two cases to be considered:

(a) If the pressure gradient is constant throughout the fluid then it can be written

$$\frac{\partial p}{\partial x} = \frac{p_2 - p_1}{l}.$$

(b) If the pressure gradient varies with x, then as the element is gradually reduced in size in the limit when $l \to 0$,

$$\frac{\partial p}{\partial x} \to \frac{p_2 - p_1}{l}.$$

Thus in either case the force on the cylinder becomes

$$-\frac{\partial p}{\partial x} \delta V.$$

The total force in the x-direction F_x, acting upon the fluid of volume V, is obtained by summing the forces on all such cylindrical elements. Hence $\partial p/\partial x$ being a constant in this integration,

$$F_x = \int -\frac{\partial p}{\partial x}\, dV$$
$$= -\frac{\partial p}{\partial x} V. \tag{9.33}$$

Similar analyses will give similar results for the components of force in the y- and z-directions.

In a fluid under uniform gravity in which the density is homogeneous, the uniform pressure gradient is given by eqn. (5.3) and so eqn. (9.33) gives

$$F_z = -\frac{\partial p}{\partial z} V$$
$$= g\varrho V$$
$$= \text{weight of the volume of fluid.}$$

The pressures around the volume, and so also the force, are independent of whether the volume contains a quantity of the fluid or whether it contains a solid body. In the latter case the result is known as the buoyancy principle of Archimedes.

There is a second facet to Archimedes principle: if the volume contains fluid then it is obvious that for a balance of moments against a weight force the pressure force must in addition act through the centre of gravity of the volume which, for the uniform gravity specified, is the same as the centre of mass.

In the example of the atmosphere that was considered in § 9.7, it was noted that the sun heats the earth's surface at a much greater rate than conduction can apply heat upwards. Thus at the surface the internal energy, by eqn. (6.9), and hence the temperature, by eqn. (9.31), is increasing with time. It follows from eqn. (9.11) that, the pressure remaining constant, the density drops.

If this happens to a localized finite quantity of air its volume increases and so, by eqn. (9.33), does the pressure force exerted by the surrounding atmosphere. This results in an out-of-balance force upon this volume of air causing it to move vertically upwards. This vertical motion, known as natural convection,[†] carries warm air upwards and reduces the temperature drop to below the value calculated in § 9.7.

Natural convection also occurs in liquids, being particularly noticeable in a liquid that is being heated from below.

References

1. KAYE, G. W. C. and LABY, T. H., *Tables of Physical and Chemical Constants*, 9th edn., Longmans, London, 1941, p. 65.
2. ZEMANSKY, M. W., *Heat and Thermodynamics*, 4th edn., McGraw-Hill, New York, 1957, table 13.1.
3. BICKLEY, W. G., Formulae for numerical integration, *The Math. Gazette* **23** (256), 352 (Oct. 1939).
4. HILSENRATH, J. et al., *Tables of Thermal Properties of Gases*, NBS Circular 564, 1960.
5. *Idem*, New values for the physical constants, *Nat. Bur. Stand. Tech. News Bull.* **47** (10), 175 (1963).
6. WILSON, J. L. and REGAN, J. D., *Calculations of the Thermodynamic Properties* of *Nitrogen at High Pressures*, NPL Aero. Rep. 1089, Jan. 1964.
7. VAN DER WAALS, J. D., *The Continuity of the Liquid and Gaseous States: Physical Memoirs*, Physical Society of London, vol. 1, pt. 3, p. 332 (1888–90).
8. KAYE, G. W. C. and LABY, T. H., *Tables of Physical and Chemical Constants*, 9th edn., Longmans, London, 1941, p. 43.
9. KUBIN, R. F. and PRESLEY, L. L., *Thermodynamic Properties and Mollier Chart for Hydrogen from $300°K$ to $20,000°K$*, NASA SP-3002, 1964.

---

† Aeroplane pilots—in particular sailplane pilots—and meteorologists refer to this as thermal motion.

10. ZEMANSKY, M. W., *Heat and Thermodynamics*, 4th edn., McGraw-Hill, New York, 1957, art. 9.7.
11. KAYE, G. W. C. and LABY, T. H., *Tables of Physical and Chemical Constants*, 9th edn., Longmans, London, 1941, p. 77.
12. ZEMANSKY, M. W., *Heat and Thermodynamics*, 4th edn., McGraw-Hill, New York, 1957, art. 9.7.

CHAPTER 10

MIXTURES OF PHASES

10.1 Phase distinction

The word phase implies that a change in the state of a substance can be observed by a change in its appearance. Later it will be seen that this is not always possible.

Three phases are distinguished here—the solid, the liquid, and the gaseous phases. The term fluid, by the definition given in § 5.3, describes substances in either the liquid or gaseous phase.

The surfaces dividing the different phases in a mixture can sometimes be observed optically by the reflection of light as from a solid–gas interface and from a liquid–gas interface and, for transparent materials, by the refraction of light at a solid–liquid interface.

10.2 The liquid–gas boundary

At a liquid–gas interface a meniscus exists having a surface tension as defined in § 2.3. The surface tension is found to be a property that is mainly a function of the temperature. However, on very small drops of liquid the surface tension is found to be affected by the radius of curvature.[1] Also very small droplets are usually electrically charged to a degree that has a significant effect upon the surface tension.[2]

When a liquid–gas interface joins a solid wall, the angle between the two is only rarely a right angle. Two examples are illustrated in Fig. 10.1. They are the cases of mercury and water each contained within a vertical glass tube and each in contact with air. The former has a meniscus that is convex upwards, the latter one that is concave upwards. The ranges of contact angle quoted in this figure are caused by the varying degrees of chemical contamination of the liquid and glass surfaces.

When the contact angle is not 90°, the surface tension v exerts a force along the axis of the tube as shown sketched in

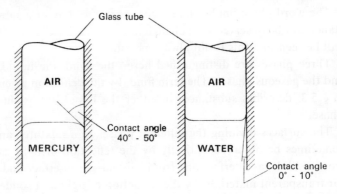

Fig. 10.1

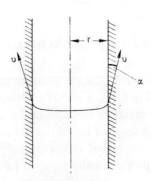

Fig. 10.2

MIXTURES OF PHASES

Fig. 10.2. From the definition of eqn. (2.2) this force is the product of the surface tension and the length of the line along which it is acting. The line length is $2\pi r$, where r is the tube radius.

The component of the surface tension in the axial direction is $v \cos \alpha$, and so the force in this direction is $2\pi r v \cos \alpha$. The area of the tube being πr^2, then this force is equivalent to an added pressure of

$$\frac{2\pi r v \cos \alpha}{\pi r^2} = \frac{2v \cos \alpha}{r}.$$

As sketched in Fig. 10.2, this equivalent pressure acts upwards upon the liquid surface and so raises the liquid level against gravity by an amount z. In accordance with eqn. (9.20) this equivalent pressure equals $g\varrho z$.

Equating these two quantities gives

$$g\varrho z = \frac{2v \cos \alpha}{r}$$

or

$$z = \frac{2v \cos \alpha}{g\varrho r}. \tag{10.1}$$

For water in contact with glass at 15° C,

$$\begin{aligned}
v &= 74 \text{ dynes cm}^{-1} \\
&\triangleq \text{g sec}^{-2} \\
&= 74 \times 10^{-3} \text{ kg sec}^{-2} \\
&= \frac{74 \times 10^{-3}}{0.454 \times 32.2} \text{ slug sec}^{-2} \\
&= 5 \times 10^{-3} \text{ slug sec}^{-2} \\
&\triangleq \text{lb ft}^{-1}
\end{aligned}$$

Also the density of water is 1·94 slug ft^{-3}, and taking α as 10°, then eqn. (10.1) gives

$$zr = \frac{2 \times 5 \times 10^{-3} \times 0.98}{32.2 \times 1.94}$$
$$= 1.6 \times 10^{-4} \text{ ft}^2.$$

In a manometer in which a height of liquid is used as a measure of pressure, this height increment z will alter the reading.[†] Limiting z to 0·01 in., then,

$$r = \frac{1.6 \times 10^{-4} \times 12}{0.01}$$
$$= 0.19 \text{ ft}$$
$$= 2.3 \text{ in.,}$$

which implies an inconveniently large manometer tube. If, however, such a manometer is being used to measure the difference in height between two successive positions of the meniscus, then the effect of surface tension cancels provided the tube bore is precisely constant along the tube axis. Still using z to denote the height increment due to surface tension, then from eqn. (10.1)

$$dz = -\frac{2v \cos \alpha}{g\varrho r^2} dr.$$

If the value of z along the tube is not to vary by more than 0·01 in., and if the fractional change in radius dr/r is to be limited to 0·05, then substituting into the above equation, and noting that if dr/r is positive then dz is negative,

$$r = \frac{1.6 \times 10^{-4} \times 0.05 \times 12}{0.01}$$
$$= 0.96 \times 10^{-2} \text{ ft}$$
$$= 0.115 \text{ in.,}$$

a much more reasonable design criterion.

[†] For a description of such manometers, see P. Bradshaw, *Experimental Fluid Mechanics*, Pergamon, 1964, in the same series.

MIXTURES OF PHASES

Surface tension can have a marked effect upon the state of the liquid contained by the meniscus. This can happen in the case of a small sphere of liquid, a sector of which is shown sketched in Fig. 10·3. Upon this portion of meniscus, the inward component of the surface tension plus the component of the atmospheric pressure is balanced by the internal pressure of

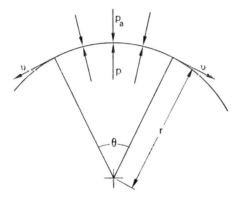

Fig. 10.3

the liquid. The inward component of the surface tension acting along a line of length $2\pi r \sin \theta/2$ gives a force of,

$$v \sin \frac{\theta}{2} 2\pi r \sin \frac{\theta}{2}.$$

Using the result of eqn. (9.32), the force due to the atmospheric pressure p_a, in the same direction, is

$$p_a \pi \left(r \sin \frac{\theta}{2} \right)^2.$$

and similarly the outward force due to the internal pressure p is

$$p\pi \left(r \sin \frac{\theta}{2} \right)^2.$$

Equating these forces for balance gives

$$v2\pi r \sin^2 \frac{\theta}{2} + p_a \pi r^2 \sin^2 \frac{\theta}{2} = p\pi r^2 \sin^2 \frac{\theta}{2}$$

or

$$p = p_a + \frac{2v}{r}.$$

With a sphere diameter of 1 μ, a lower value not uncommon in liquid sprays, then

$$r = \frac{1}{2} \times 10^{-4} \text{ cm}$$

$$= \frac{10^{-4}}{2 \times 2 \cdot 54 \times 12} \text{ ft,}$$

so that

$$\frac{2v}{r} = \frac{2 \times 5 \times 10^{-3} \times 2 \times 2 \cdot 54 \times 12}{10^{-4}}$$

$$= 6 \cdot 1 \times 10^3 \text{ lb ft}^{-2}$$

$$= 2 \cdot 9 \text{ atm.}$$

Thus with an external pressure of 1·0 atm the internal pressure is 3·9 atm.

10.3 Phase changes

The three phase changes of H_2O are known as ice, water, and steam.[†]

If 1 kg of ice has heat applied to it, its internal energy will increase. If this process is performed at atmospheric pressure, then the temperature of the ice will rise with the internal energy

[†] Some writers use the term water for any phase of H_2O and some distinguish water by calling it liquid water. If H_2O was here called hydrogen-oxide no one would know what was being referred to.

increase. This is illustrated in Fig. 10.4, for a series of equilibrium states, by the solid line AB.[†] At $0°C$ the internal energy would increase by about 0.33×10^6 J, a quantity here called the fusion energy, whilst the ice was gradually being converted into water. This is indicated by the line marked BC, which forms a phase boundary between the solid and liquid phases.

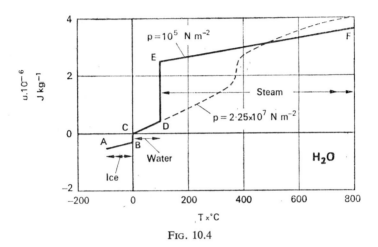

Fig. 10.4

On completion of this conversion the internal energy would continue to rise following the line CD until the temperature reached $100°C$. Again there would be a pause in the temperature rise whilst the water was converted to steam, and the internal energy rose by about 2.1×10^6 J, a quantity here called the vaporization energy. This part of the process is indicated by the line DE which forms another phase boundary between liquid and gas. Further application of heat would result in successive

[†] The change in U cannot be equated to Q as the volume of the ice would change so that work would be done against the atmospheric pressure and the amount of this work would depend upon the form of the process. See also § 16.5.

equilibrium states being represented by points along the line EF.

Four matters arise from this form of the variation of u with T. First, whereas it has been noted that in the case of either liquids or solids alone, the internal energy could change with change in pressure, the temperature remaining constant, here at the phase boundaries the internal energy of the mixture changes even though both the pressure and temperature are remaining constant.

Second, there is a mathematical discontinuity in u across the phase boundaries BC and DE. Thus for changes of phase, du the perfect differential does not exist. This is analogous to the precipice example in the discussion of height in § 1.1.

Third, if du does not exist at the phase boundaries then C_u, which is based upon a differentiation of u, mathematically does not exist. Approaching $0°C$ from above, the value of C_u approaches a value that is about double that obtained if this temperature is approached from below.

And fourth, the pressure and the temperature together do not define the state at a phase boundary. At each of these boundaries the internal energy seems to have an infinity of values. However, there are also an infinity of states. For instance at B the state is one of completely ice, at C it is one of completely water; in between these points corresponds to a mixture of ice and water in varying proportions.

By specifying the pressure and temperature corresponding to the phase boundary BC the internal energy per mass unit is known for both the solid ice and the liquid water separately. These quantities are denoted by u_s and u_l. Considering a mixture of ice and water at some equilibrium state along BC that has a mass of 1 slug, then, if the mass of ice is r slugs, the mass of water is $(1-r)$ slugs. The internal energy of the solid content of the mixture U_s is given by

$$U_s = ru_s$$

MIXTURES OF PHASES

and similarly for the liquid

$$U_l = (1-r)u_l.$$

The internal energy being an additive property then that of the mixture U is given by

$$\begin{aligned}U &= U_s + U_l \\ &= ru_s + (1-r)u_l.\end{aligned}$$

As unit mass is being considered, then

$$U = u,$$

so that, finally, the internal energy of the mixture is given by

$$u = ru_s + (1-r)u_l. \tag{10.2}$$

A similar expression can be written down for the water–steam phase boundary. In this case of a liquid–gas mixture the proportion of the mass of the gas to the mass of the mixture is called the quality.

Similar discontinuities at phase boundaries occur in graphs of the density. The mean density of a mixture can be obtained in the manner just demonstrated for the internal energy by using the fact that volume is an additive property. The volume of 1 kg of the mixture[†] is equal to $1/\varrho$ where ϱ is the mean density of the mixture. Also the volume of r kg of the solid, ice is equal to r/ϱ_s, where ϱ_s is the solid, ice density, and the volume of $(1-r)$ kg of water is equal to $(1-r)/\varrho_l$.

Thus

$$\frac{1}{\varrho} = \frac{r}{\varrho_s} + \frac{1-r}{\varrho_l}. \tag{10.3}$$

These relations will apply to substances other than H_2O which is simply adopted here as an illustrative example.

[†] The unit of mass is changed here to emphasize that the derivation of eqns. (10.2) and (10.3) is independent of the units of mass chosen.

10.4 Phase boundaries

Along the phase boundaries of a system of material which is in an equilibrium state that is stable to small disturbances the temperature is a function of the pressure only.[†] This is illustrated in Fig. 10.5 for H_2O where the three divisions of the boundary between the three phases are plotted.

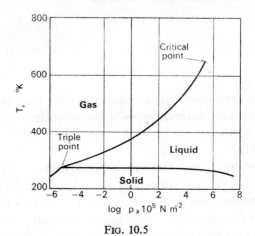

Fig. 10.5

It follows from this that the pressure and temperature alone are insufficient to fix the state at a phase boundary. For specifying a phase boundary and the pressure will give only the corresponding temperature; it will not give the proportions of the mixture. If however, the values of p and u are known, then eqn. (10.2) enables the proportion r to be calculated. Thus u and p form a pair of independent variables that fix the state of a mixture, as also do u and T; additionally, from eqn. (10.3) suitable pairs of variables are either ϱ and p or ϱ and T.

[†] The case of lack of this stability is discussed later in § 11.8.

MIXTURES OF PHASES 171

Hence a general result is that at a phase boundary the definition of the state of a two-property mixture requires that at least one of the defining properties must be additive.

10.5 Triple point

There is a point on the phase diagram of Fig. 10.5, marked the triple point, where the solid, liquid, and gaseous phases can coexist in an equilibrium state. In this case if the gas phase has an internal energy per unit mass u_g and a density ϱ_g, and if the proportions of solid, liquid, and gas are denoted by r_s, r_l, and $(1-r_s-r_l)$, then eqns. (10.2) and (10.3) are now replaced by

$$u = u_s r_s + u_l r_l + u_g(1 - r_s - r_l)$$

and
$$\frac{1}{\varrho} = \frac{r_s}{\varrho_s} + \frac{r_l}{\varrho_l} + \frac{1 - r_s - r_l}{\varrho_g}.$$

Determination of a triple-point state requires knowledge of the two proportions r_s and r_l. These two unknowns can be evaluated by the two preceding equations if ϱ and u are specified because the other terms are known for the triple point. Thus values of ϱ and u are now sufficient to fix a triple-point state and neither p nor T can be used for this purpose. A general rule is that at a triple point the definition of the state of a two-property mixture requires that both the defining properties must be additive ones.

10.6 Critical point

Referring again to Fig. 10.5, up to just over $600°$ K it is found that the gas–liquid phase boundary can be readily distinguished by the presence of a meniscus. As the temperature and pressure are raised further the meniscus disappears from view and the previously described discontinuities, in quantities like the density and the internal energy, no longer exist. This point of disappearance is called the critical point. At this point the maximum slopes of curves of quantities, such as u and ϱ, as they vary with T are infinite, but the curves are smooth; above this point the maximum slopes are finite. Such a curve for a pressure just above the critical pressure is shown dotted in Fig. 10.4.

A further point of interest can be observed from this graph. It will be noticed that it intersects the graph for the lower pressure at a temperature of about 470° K. Thus at this point the corresponding value of u, that is $3 \cdot 1$ J kg^{-1}, together with this temperature will not fix the state, for there are two possible ones corresponding to the two pressures. The independent variables now required are the pressure and the temperature.

References

1. HIRSCHFELDER, J. O., CURTISS, C. F. and BIRD, R. B., *Molecular Theory of Gases and Liquids*, Wiley, New York, 1954, p. 348.
2. STEUDEL, T. and MAYRHOFER, R. C., The influence of electric charges on interfacial energy fundamentals, *Tenside* 2 (9), 289 (Sept. 1965).

CHAPTER 11

THE CHARACTERISTICS OF FLUID MOTION

11.1 Streamlines

Using methods described elsewhere,[†] the direction of the velocity within a fluid flow could be observed at a number of points. An example of a flow is sketched in Fig. 11.1 where the velocity direction at several points is indicated by the arrows.

If these flow directions are all observed simultaneously, and if curves are then drawn which are everywhere tangential to the local flow direction, then an instantaneous picture of the flow is obtained. The curves of this picture are the streamlines of the flow pattern for the specified instant of time. Examples are drawn as dotted lines in Fig. 11.1.

From this definition of a streamline it follows that the velocity component perpendicular to a streamline is always zero.

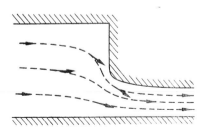

FIG. 11.1

[†] Bradshaw, *op. cit.*

This condition is the one required at a solid boundary to the flow and so a solid boundary contains streamlines.

A tube drawn in a flow whose surface is made up of streamlines is called a streamtube. Again, there will be no flow across the surface of a streamtube; when fluid is flowing through a pipe the wall forms a streamtube.

11.2 Pathlines

As a single particle of fluid travels through a flow it traces out its path. This line along which it moves is called the pathline. From the discussion of § 1.3 it follows that the direction of the velocity of this particle is tangential to the path line at all successive positions.

The difference between a streamline and a pathline is that the former gives an instantaneous picture of the velocity direction of all the particles of fluid on it; the latter gives a picture of the velocity direction of a single particle over a period of time.

11.3 Steady and unsteady flow

As an element of fluid travels along a streamline, usually its properties will change with time as will those of the elements that precede and succeed it. If all the elements have identical states at the moment that they pass a fixed point on the streamline, and if this condition applies to all points within the flow, then the flow is called a steady one. In a steady flow there are no changes with respect to time at any fixed point.

The properties concerned in this definition are the thermostatic ones such as pressure, density, and internal energy, together with mechanical properties such as velocity and acceleration.

All real flows are unsteady to some degree, but large portions of many of them can be regarded as steady to a high degree of approximation. It is not required of an unsteady flow that all properties at a point should fluctuate with respect to time; only one need do so.

A particle of fluid setting out from point A as indicated in Fig. 11.2 can have its motion specified by a relation between

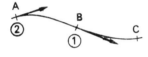

Fig. 11.2

its velocity and the time; or for the x-component this can be written

$$\dot{x} = \dot{x}(t). \tag{11.1}$$

The x-coordinate of the path is given by

$$x = \int \dot{x}\, dt,$$

and performing this integration gives x as a function of time or

$$x = x(t). \tag{11.2}$$

Similarly, the other ordinates are obtained in the form

$$y = y(t),$$
$$z = z(t),$$

and eliminating t between these last three relations gives

$$y = y(x),$$
$$z = z(x),$$

and so the shape of the path is known.

Eliminating t between eqns. (11.1) and (11.2) gives

$$\dot{x} = \dot{x}(x) \tag{11.3}$$

and so conversely, if this latter relation and similar ones for the y- and z-components are known, the analysis can be worked in reverse to give the shape of the path.

In a steady flow eqn. (11.3) is independent of time; it is the same for all the particles of fluid that pass through A and so all those particles follow the same path. When, as illustrated in the second part of Fig. 11.2, particle No. 1 has reached point B particle No. 2 is setting out from A on the same path. At any instant in time, line ABC is the path for all the particles along it; thus the velocity is everywhere tangential to ABC. But this is the definition of a streamline, and so in steady flow streamlines and path lines are identical.

It is a further consequence of a flow being steady that the states of all particles are identical at the instant when they pass through a single point. Thus, referring again to Fig. 11.2, the state of particle 1 at A is identical to the later state of particle 2 when it reaches A. The difference, then, between the successive states of particle 1 at B and at A is equal to the instantaneous difference between the state of 1 at B and that of 2 at A.

11.4 Fixed and moving axes

In analysing a fluid flow the history of an element will be traced as if by an observer travelling with the element. In experimental work, measurements are made at a fixed point as the fluid flows by. A means of conversion from such fixed to stationary axes is now developed.

Figure 11.3 is a sketch of a closed heat-insulated pipe circuit around which water is circulated by a pump and heated by a heating coil. Along one path line are three temperature measuring points denoted 1, 2, and 3. The element of fluid at position 3, having been subjected to heating as it moved along the pathline, would be of higher temperature than those following it and which are at stations 1 and 2. If at one instant of time, t_α say, the three temperatures are measured and denoted by T_1, T_2, and T_3, a distribution along the path line such as that

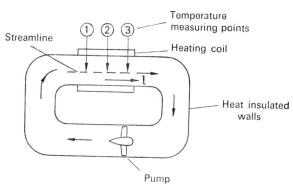

Fig. 11.3

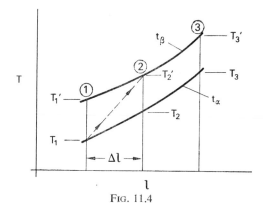

Fig. 11.4

sketched in Fig. 11.4 might result. It should be noted that these three temperatures are those of the three different elements of the fluid that happen to be at the measuring positions at this instant of time. The heating process being continued, a set of readings at a later time t_β, say, would show temperatures of a generally higher level giving the curve sketched with temperatures T'_1, T'_2, and T'_3, say.

The time interval $(t_\beta - t_\alpha)$ can be so chosen that during it the particle of fluid which had a temperature T_1 at station 1 at time t_α would have travelled to station 2, where it would then have a temperature of T'_2: and this temperature would not be T_2 which would have been the prior temperature of another particle now travelling ahead.

Thus the temperature change ΔT of the element of fluid as it moves from position 1 to position 2 is given by

$$\Delta T = T'_2 - T_1,$$

and this change has taken place in the time interval Δt, where

$$\Delta t = t_\beta - t_\alpha.$$

In the limit, as $\Delta t \to 0$, the rate of change of temperature of the particle, written DT/Dt, is given by

$$\frac{DT}{Dt} = \underset{t_\beta \to t_\alpha}{Lt} \frac{T'_2 - T_1}{t_\beta - t_\alpha} = \underset{t_\beta \to t_\alpha}{Lt} \frac{\Delta T}{\Delta t}. \qquad (11.4)$$

The increment ΔT has two portions, that is

$$\Delta T = (T'_2 - T_2) + (T_2 - T_1)$$
$$= \frac{T'_2 - T_2}{\Delta t} \Delta t + \frac{T_2 - T_1}{\Delta l} \Delta l,$$

where Δl is the distance between stations 1 and 2. In the limit, as $\Delta t \to 0$, the first term is written

$$\frac{T'_2 - T_2}{\Delta t} \Delta t = \frac{\partial T}{\partial t} \Delta t,$$

THE CHARACTERISTICS OF FLUID MOTION   179

the differential being the rate of change of temperature with time at a fixed position.

Similarly, as $\Delta l \to 0$, the second term is written

$$\frac{T_2 - T_1}{\Delta l} \Delta l = \frac{\partial T}{\partial l} \Delta l,$$

and this differential is the rate of change with distance at a fixed instant of time. Thus eqn. (11.4) can be rewritten

$$\frac{DT}{Dt} = Lt \left[\frac{1}{\Delta t} \left(\frac{\partial T}{\partial t} \Delta t + \frac{\partial T}{\partial l} \Delta l \right) \right]$$
$$= Lt \left[\frac{\partial T}{\partial t} + \frac{\partial T}{\partial l} \frac{\Delta l}{\Delta t} \right].$$

The limit now corresponds to both $\Delta t \to 0$ and $\Delta l \to 0$. In this limit, $\Delta l / \Delta t = q$, the velocity of the fluid element, and so

$$\frac{DT}{Dt} = \frac{\partial T}{\partial t} + q \frac{\partial T}{\partial l}. \tag{11.5}$$

This result is generally applicable to other properties of the fluid element. If the pressure p is considered, then

$$\frac{Dp}{Dt} = \frac{\partial p}{\partial t} + q \frac{\partial p}{\partial l}$$

or, if the velocity q is observed, then

$$\frac{Dq}{Dt} = \frac{\partial q}{\partial t} + q \frac{\partial q}{\partial l}, \tag{11.6}$$

where Dq/Dt is the acceleration of the element.

By definition, in a steady flow all the partial differentials with respect to time are zero. That is, in these three examples,

$$\frac{\partial T}{\partial t} = 0, \quad \frac{\partial p}{\partial t} = 0, \quad \frac{\partial q}{\partial t} = 0.$$

11.5 Two-dimensional flow

All real flows require three coordinates, x, y, and z, say, to fix the position of a particle of fluid, and in such flows each particle in general has a resultant velocity q which comprises the three components $\dot{x}, \dot{y}$, and $\dot{z}$. A two-dimensional flow is an artificial concept to which many real flows closely approximate. An example is the flow along a long parallel duct of rectangular cross-section across which there is a slender strut; this is illustrated in Fig. 11.5. If the z-axis is drawn parallel to the strut axis, then to a close approximation only x- and y-ordinates are required to describe the major part of the flow in which only the $\dot{x}$ and $\dot{y}$ velocity components exist, $\dot{z}$ being zero. The flow pattern is the same in all x–y planes.

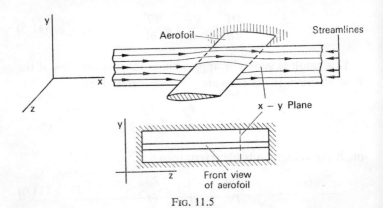

Fig. 11.5

An element of fluid, as it moves along a streamline, can both rotate and distort in shape. This is illustrated for a two-dimensional flow in the two sketches of Fig. 11.6; both illustrate the motion of a triangular element ABC along a streamline to a new position $A'B'C'$. The first case is one of rotation

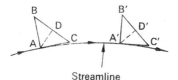

Streamline

a) Rotation without distortion

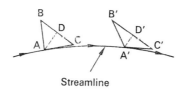

Streamline

b) Distortion without rotation

Fig. 11.6

without any distortion of the shape, the second is of distortion without any rotation. The rotation can be indicated by the direction of the mean line AD; in the second case of no rotation, $A'D'$ remains parallel to AD.

11.6 Shear stress in a moving fluid

It has been pointed out in § 5.3 that when a Newtonian fluid flows past a solid boundary the velocity at the boundary is zero and the shear stress τ is related to the velocity gradient by eqn. (5.1).

A small element of fluid adjacent to a solid boundary is shown sketched in Fig. 11.7. As a result of a two-dimensional motion the boundary exerts an equal and opposite reaction on the element giving a shear stress τ_0 in the direction shown.

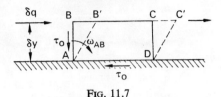

Fig. 11.7

The angular velocity of AB, ω_{AB}, to an observer sitting on AD is

$$\omega_{AB} = \frac{\delta q}{\delta y}$$

which for an infinitesimal element becomes

$$\omega_{AB} = \left(\frac{\partial q}{\partial y}\right)_{y=0}.$$

Thus eqn. (5.1) can be expressed

$$\tau_0 = \mu \omega_{AB}. \tag{11.7}$$

But an observer sitting on AB sees that AD has an equal and opposite angular velocity ω_{AD}. It follows that, as viscosity is a molecular effect having no preferred direction in a Newtonian fluid (§ 5.3), a shear stress must similarly act along AB and, furthermore, it will be given by

$$\tau = \mu \omega_{AD}$$
$$= -\mu \omega_{AB}$$
$$= -\tau_0,$$

and its direction will be as shown sketched.

Experiments indicate that eqn. (11.7) applies at any instant throughout the flow where now ω_{AB} in that equation is the instantaneous angular velocity of AB relative to AD. In the case of unsteady flows, this statement is further amplified in § 11.12.

Figure 11.8 is a sketch of a triangular element of fluid *ABC* in motion along a streamline. In this sketch *r* indicates the radius of curvature of the streamline at *A*, *q* the velocity of *A*, $q+\delta q$ that of *B*, and δn the length of *AB* normal to the

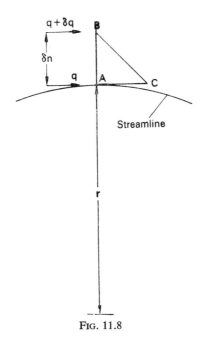

Fig. 11.8

streamline. The angular velocity of *AC*, ω_{AC}, is given by

$$\omega_{AC} = q/r. \tag{11.8}$$

That of *AB* is

$$\omega_{AB} = \frac{(q+\delta q)-q}{\delta n}$$

$$= \frac{\partial q}{\partial n} \tag{11.9}$$

The angular velocity of AB relative to AC, ω, is thus

$$\omega = \omega_{AB} - \omega_{AC}$$
$$= \frac{\partial q}{\partial n} - \frac{q}{r}.$$

Then the shear stress at A acting along both AB and AC is given, from eqn. (11.7), by

$$\tau = \mu \left[\frac{\partial q}{\partial n} - \frac{q}{r} \right], \qquad (11.10)$$

and at a curved solid boundary, at which $q = 0$,

$$\tau_0 = \mu \left(\frac{\partial q}{\partial n} \right)_0. \qquad (11.11)$$

It is recalled that these results only apply for Newtonian fluids.

11.7 Comparison of normal and tangential stresses

A typical value of the normal stress in a gas is sea-level atmospheric pressure, that is $14 \cdot 7$ lb in^{-2}. A typical velocity gradient at a solid boundary is 10^4 sec^{-1}. The viscosity of air at 23°C is

$1 \cdot 83 \times 10^{-4}$ gm cm^{-1} sec^{-1} [†]

$$= \frac{1 \cdot 83 \times 10^{-4} \times 10^{-3} \times 2 \cdot 205 \times 10^2 \times 0 \cdot 914}{32 \cdot 2 \times 3} \text{ slugs ft}^{-1} \text{ sec}^{-1}$$

$= 3 \cdot 82 \times 10^{-7}$ slugs ft^{-1} sec^{-1}.

Thus a typical value of τ is $3 \cdot 82 \times 10^{-7} \times 10^4 = 3 \cdot 82 \times 10^{-3}$ lb ft^{-2} to be compared with the typical normal stress of $14 \cdot 7 \times 144 = 2 \cdot 12 \times 10^3$ lb ft^{-2}. There is a factor of 10^6 in a comparison of these two stresses. And yet, whilst

[†] This unit is called the poise,

shear stresses are usually very small compared with normal stresses, by no means can they always be neglected. This statement should become acceptable as discussion develops.

11.8 Acceleration effects

The many characteristics of a fluid, such as its pressure, temperature, viscosity, surface tension, internal energy, thermal conductivity, and so on, derive from the motion and interactions of its molecules. A change in, for instance, the pressure results in a change in the molecular motion, and this change must take time. In air at normal atmospheric conditions the number of collisions a molecule makes is of the order of 10^9 per sec. And so, if a new continuum equilibrium condition is achieved after about 10 collisions this takes place in 10^{-8} sec. The frequency of collisions, at a constant temperature, is proportional to the pressure, and so at extremely low pressures the time for equilibrium to be achieved can be sufficient to provide a noticeable time lag in a rapidly accelerating flow.[1] During this period of lack of equilibrium the previous definitions of pressure and temperature are no longer valid as they suppose equilibrium conditions.

Other effects can be significant in flows where the pressures are of the more commonly occurring level. For instance, in a flow of steam which is condensing, the phase change can be delayed beyond the equilibrium phase boundary corresponding to the local flow pressure.[2] The state of the steam is not a stable equilibrium one. A disturbance results in a spontaneous change of phase. For a fixed pair of properties such as internal energy and pressure there are then two possible states; one of these is a stable equilibrium one of a mixture of steam and water, the other an unstable state of only steam. When the change does take place it occurs rapidly, introducing a marked effect into the flow. Similar effects occur in the condensation of the moisture contained in air expanding rapidly through a nozzle.[3]

Study of such effects is left to further reading to include the papers cited.

11.9 Pressure in a moving fluid

As all real flows have velocity gradients normal to streamlines, they have a distribution of shear stress within the flow. Thus the result demonstrated in § 2.5 that the normal stress is the same in all directions does not apply in general throughout a fluid in motion because it depends upon an absence of shear stresses.

An attempt to define a pressure at a point in a moving fluid can be approached in the following three ways provided that the previously described time-lag effects are absent.

First, the fact that the tangential stress is very small compared with the normal stress can be used to justify ignoring the variation of normal stress with direction.

Second, it can be shown that the mean value of the three normal stresses along three mutually perpendicular axes is independent of the orientation of those axes.[4] This mean value of the stresses can be adopted as the value of the pressure.

And third, by combining eqn. (9.11) with eqn. (9.31), the internal energy and the pressure can be related by

$$p = \frac{R\varrho u}{MC_u} \qquad (11.12)$$

for a gas in equilibrium. As the value of ϱ for a gas is the mass of the molecules per volume unit, and as the internal energy can be identified with the sum of the molecular kinetic energies per volume unit, these two quantities can be satisfactorily defined in this way even when motion is present. Thus pressure

THE CHARACTERISTICS OF FLUID MOTION

can be defined so that eqn. (11.12) is satisfied exactly for a gas in motion.[5]

These three approaches to a definition of a pressure for a fluid in motion have relative advantages and disadvantages, and the validity of using the pressures so defined in the equation of state for a gas in equilibrium still raises uncertainties.[6] Further stwely is here deferred. In the elementary discussion of this volume the first approach is accepted as satisfactory.

Often, in discussing a fluid flow, the pressure is referred to as the static pressure; the adjective is superfluous and is not used here.

11.10 Rotation in a moving fluid

The bouyancy principle of Archimedes was described in § 9.11. In particular, for an infinitesimal element of an incompressible fluid under a pressure gradient, the resultant of the pressure stresses acts along a line through the centre of mass of the volume. Thus this important result is obtained for an incompressible flow; the force resulting from the normal stresses acting upon an infinitesimal element of a fluid in motion can never cause the element to rotate. If an element of fluid rotates as it moves along a path line then this rotation can only be initiated by tangential, viscous stresses. It does not necessarily follow that viscous stresses will cause rotation.

However, in a compressible flow the angular velocity can be changed by the pressure force. This is illustrated by the simple case, sketched in Fig. 11.9, of an element of fluid acted upon by the force due solely to a pressure gradient $\partial p/\partial y$. According to the principle of Archimedes this force will act through the centre of volume of the element; that is, if the element is of length x, and of unit thickness perpendicular to the x–y plane, this force will act at a position at $x/2$ along the element.

If the flow is compressible a variation of density can occur

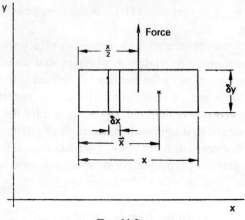

FIG. 11.9

along the x-axis. Writing this variation in the form

$$\varrho = a + bx, \qquad (11.13)$$

then the mass of the element δm is given by

$$\delta m = \delta y \int_0^x (a+bx)\, dx$$

$$= \delta y\, x(a + \tfrac{1}{2} bx).$$

Then taking moments gives the position of the centre of mass of the element $\bar{x}$ as

$$\bar{x} = \frac{\delta y \int_0^x x(a+bx)\, dx}{\delta y\, x(a + \tfrac{1}{2} bx)}$$

$$= x\, \frac{\tfrac{1}{2} a + \tfrac{1}{3} bx}{a + \tfrac{1}{2} bx}. \qquad (11.14)$$

THE CHARACTERISTICS OF FLUID MOTION

The force on the element will act at a distance of $\bar{x}-(x/2)$ from the centre of mass which is thus given by

$$\bar{x}-\frac{x}{2} = x\left(\frac{\frac{1}{12}bx}{a+\frac{1}{2}bx}\right).$$

Shrinking the element to a square one of length, $x = \delta y$, makes this expression for the moment arm equal to

$$\frac{1}{12}\frac{b(\delta y)^2}{a+\frac{1}{2}b(\delta y)}.$$

When only first-order terms are retained, so that $\frac{1}{2}b(\delta y)$ is negligible compared with a this becomes

$$\frac{1}{12}\frac{b}{a}(\delta y)^2.$$

The volume of the element being $(\delta y)^2$, then the force upon it is given by eqn. (9.33) as

$$-\frac{\partial p}{\partial y}(\delta y)^2.$$

The moment of inertia, about its centre, of a vertical strip of the element is

$$\varrho\,\delta x\,\delta y\,\frac{(\delta y)^2}{12}.$$

The moment of inertia of this strip about the centre of mass of the element is then

$$\varrho\,\delta x\,\delta y\,\frac{(\delta y)^2}{12}+\varrho\,\delta x\,\delta y(x-\bar{x})^2.$$

Thus the moment of inertia of the element about its centre of mass is

$$\int_0^x\left[\varrho\,\delta x\,\delta y\,\frac{(\delta y)^2}{12}+\varrho\,\delta x\,\delta y(x-\bar{x})^2\right],$$

which, with substitution from eqn. (11.13), is

$$\delta y \int_0^x (a+bx)\left[\frac{(\delta y)^2}{12}+(x-\bar{x})^2\right]dx$$

$$= \delta y\left[\frac{(\delta y)^2}{12}\left(ax+\frac{1}{2}bx^2\right)+\frac{1}{3}a(x-\bar{x})^3+\frac{1}{3}a\bar{x}^3 \right.$$

$$\left. +b\left(\frac{x^4}{4}-\frac{2}{3}\bar{x}x^3+\frac{1}{2}\bar{x}x^2\right)\right].$$

Again, putting $x = \delta y$ and substituting for $\bar{x}$ from eqn. (11.14), this expression contains terms of order $(\delta y)^4$ and of order $(\delta y)^5$. Neglecting the latter, then the moment of inertia becomes

$$\delta y\left[\frac{(\delta y)^2}{12}a\,\delta y+\frac{1}{3}a\left(\frac{\delta y}{2}\right)^3+\frac{1}{3}a\left(\frac{\delta y}{2}\right)^3\right] = \frac{1}{6}a(\delta y)^4.$$

Using these results the angular acceleration $\dot{\omega}$ is given by

$$\dot{\omega} = \frac{-\dfrac{\partial p}{\partial y}(\delta y)^2 \dfrac{1}{12}\dfrac{b}{a}(\delta y)^2}{\dfrac{1}{6}a(\delta y)^4}$$

$$= -\frac{1}{2}\frac{\partial p}{\partial y}\frac{b}{a^2}.$$

Inspection of eqn. (11.13) shows that for this infinitesimal element

$$a = \varrho \quad \text{and} \quad b = \frac{\partial \varrho}{\partial x}.$$

Thus

$$\dot{\omega} = -\frac{1}{2}\frac{1}{\varrho^2}\frac{\partial p}{\partial y}\frac{\partial \varrho}{\partial x}$$

$$= \frac{1}{2}\frac{\partial p}{\partial y}\frac{\partial}{\partial x}\left(\frac{1}{\varrho}\right).$$

THE CHARACTERISTICS OF FLUID MOTION

So if there is a density gradient that is not aligned with the pressure gradient, then a change in the angular velocity of an element will occur. If the fluid is a gas obeying eqn. (9.11), then the pressure being constant in the x-direction, the density gradient will correspond to a temperature gradient and so heat will, by eqn. (7.1), be acting in the x-direction. Thus heat can be the cause of a change in the angular velocity just as friction can.

Body forces which are proportional to the mass, such as weight when g is a constant, cannot cause rotation because they will act through the centre of mass. But body forces such as magnetic or electric forces can cause rotation. If, for example, an incompressible fluid containing a distribution of electric charges has a gradient of charge concentration at an angle to the electric field direction, then the electrical force on an element will be offset from the latter's centre of mass.

The angular rotation of an element of fluid has already been illustrated in Fig. 11.6. Referring again to this figure the mean angular velocity of the element ω is thus

$$\omega = \tfrac{1}{2}(\omega_{AB}+\omega_{AC}),$$

and from eqns. (11.8) and (11.9)

$$\omega = \frac{1}{2}\left(\frac{q}{r}+\frac{\partial q}{\partial n}\right). \tag{11.15}$$

Angular velocities can be quite high. In the numerical example of § 11.7 $\partial q/\partial n = 10^4 \text{ sec}^{-1}$. If this occurs along a flat surface for which $r = \infty$, then $\omega = 0.5\times 10^4 \text{ sec}^{-1} = 0.5\times 10^5 \text{ rev/min}$. This very high angular velocity has been developed by the very small viscous stresses.

11.11 The vortex

A steady two-dimensional flow for which all the streamlines are concentric circles is called a vortex flow.[†] Such a flow is sketched in Fig. 11.10 where the fluid is shown to be moving in circles about the vortex centre O. In this flow there is no change in the properties with the angular ordinate α, all varying proper-

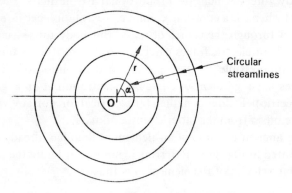

Fig. 11.10

ties are functions only of the radial ordinate r. In particular, the velocity q and the angular velocity ω are constant along a streamline. The latter fact implies that any viscous stresses acting are not contributing a torque resulting in an angular acceleration. This point is now investigated.

A sketch of an element of this flow is given in Fig. 11.11 where the shear stresses are indicated. Along AD the stress is τ, along BC, $\tau + \delta\tau$, and, from the discussion of § 11.6, along AB and DC the stress varies from τ to $\tau + \delta\tau$, having the mean value of $\tau + (\delta\tau/2)$.

[†] A vortex can more generally be a three-dimensional flow by having a curved axis; the smoke ring is an example.

THE CHARACTERISTICS OF FLUID MOTION

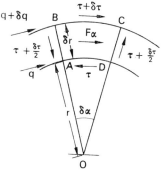

Fig. 11.11

The torque N_z about a z-axis through O is

$$N_z = (\tau+\delta\tau)(r+\delta r)\,\delta\alpha(r+\delta r) - \tau r\,\delta\alpha\,r$$
$$= (2\tau\,\delta r + r\,\delta\tau)r\,\delta\alpha$$

The moment of inertia about O, is $\varrho r\,\delta\alpha\,\delta r\,r^2$ and so the angular acceleration $\dot\omega$ is

$$\dot\omega = \frac{1}{\varrho r}\left(2\frac{\tau}{r} + \frac{d\tau}{dr}\right).$$

Then, if $\dot\omega = 0$,

$$2\frac{\tau}{r} + \frac{d\tau}{dr} = 0. \tag{11.16}$$

Substituting from eqn. (11.10) and noting that now n is in the r-direction, this result becomes[†]

$$\frac{2}{r}\mu\left(\frac{dq}{dr} - \frac{q}{r}\right) + \mu\left(\frac{d^2q}{dr^2} - \frac{1}{r}\frac{dq}{dr} + \frac{q}{r^2}\right) = 0$$

or

$$\frac{d^2q}{dr^2} + \frac{1}{r}\frac{dq}{dr} - \frac{q}{r^2} = 0.$$

[†] This result can be obtained from a solution of the full equations of viscous motion.[7]

The solution of this equation is

$$q = Ar + \frac{B}{r},$$

where A and B are constants.

There are two cases of interest derivable from this solution. First, there is the case that q must not be infinite at $r = 0$. This means $B = 0$ and

$$q = Ar. \tag{11.17}$$

The fluid then rotates as a solid body with no relative motion within it and with an angular velocity A. This value of the angular velocity is confirmed by substitution of eqn. (11.17) into eqn. (11.15), giving

$$\omega = \tfrac{1}{2}(A+A) = A.$$

This type of flow is called a forced vortex.

Second, there is the case that q must not be infinite at $r = \infty$. This means that $A = 0$ so that

$$q = \frac{B}{r} \tag{11.18}$$

and the velocity drops off hyperbolically as r increases. This type of flow is called a free vortex.

The angular velocity of an element in the free vortex is given by substituting eqn. (11.18) into (11.15) resulting in

$$\omega = \frac{1}{2}\left(\frac{B}{r^2} - \frac{B}{r^2}\right) = 0.$$

This type of flow in which the angular velocity of an infinitesimal element is zero, occurs commonly and is of considerable importance as being reasonably amenable to analysis. It is known as an irrotational flow.

Real isolated vortices, free of any solid surface, are a combination of a forced vortex surrounded concentrically by a free vortex.

THE CHARACTERISTICS OF FLUID MOTION

It is of further interest to consider the linear acceleration along a streamline. The force in the circumferential direction F_α is, by reference to Fig. 11.11,

$$F_\alpha = (\tau + \delta\tau)(r + \delta r)\,\delta\alpha - \tau r\,\delta\alpha + 2\left(\tau + \frac{\delta\tau}{2}\right)\delta r\,\frac{\delta\alpha}{2}$$
$$= (2\tau\,\delta r + r\,\delta\tau)\,\delta\alpha.$$

The mass of the element is $\varrho r\,\delta\alpha\,\delta r$ and so the acceleration in the circumferential direction $\dot{q}_\alpha$ due only to viscous shear stresses is

$$\dot{q}_\alpha = \frac{F_\alpha}{\varrho r\,\delta\alpha\,\delta r} = \frac{1}{\varrho}\left(2\,\frac{\tau}{r} + \frac{d\tau}{dr}\right).$$

Thus by eqn. (11.16) for this type of flow

$$\dot{q}_\alpha = 0, \qquad (11.19)$$

and this result is valid for both the free and forced vortex solutions.

For the forced vortex, use of the solution of eqn. (11.17) in eqn. (11.10) gives

$$\frac{\tau}{\mu} = A - A = 0.$$

This result, that the shear stresses within a forced vortex are zero, is as might be expected because there is no relative motion between the elements. Thus for a forced vortex the result of eqn. (11.19) might be expected.

For the free vortex, eqns. (11.18) and (11.10) give

$$\frac{\tau}{\mu} = -\frac{B}{r^2} - \frac{B}{r^2} = -\frac{2B}{r^2} \qquad (11.20)$$

and

$$\frac{1}{\mu}\,\frac{d\tau}{dr} = \frac{4B}{r^3}. \qquad (11.21)$$

The free vortex is the more interesting case because it is now seen that the acceleration due to the viscous shear stresses is zero even though, by eqn. (11.20), the stresses exist. This is a consequence of the rotation ω being zero and is a result that is generally applicable.

The rate at which work is done on the element by the viscous shear stresses is also of interest. Referring again to Fig. 11.11 the rate of doing work $\dot{W}$ is

$$\dot{W} = (\tau + \delta\tau)(r + \delta r)\,\delta\alpha(q + \delta q) - \tau r\,\delta\alpha\,q,$$

it being noted that the velocity being perpendicular in direction to AB and CD, no work is done by the stresses on these surfaces. Then

$$\dot{W} = [(r\,\delta\tau + \tau\,\delta r)q + \tau r\,\delta q]\,\delta\alpha.$$

The mass of the element δm is

$$\delta m = \varrho r\,\delta\alpha\,\delta r,$$

and so the rate of doing work per mass unit is

$$\frac{\dot{W}}{\delta m} = \frac{1}{\varrho}\left[\left(\frac{d\tau}{dr} + \frac{\tau}{r}\right)q + \tau\frac{dq}{dr}\right]. \qquad (11.22)$$

For a forced vortex, in which $\tau = 0$, then $\dot{W}/\delta m = 0$, but for a free vortex substitution of eqns. (11.18), (11.20) and (11.21) into eqn. (11.22) results in,

$$\frac{\dot{W}}{\delta m} = \frac{\mu}{\varrho}\left[\left(\frac{4B}{r^3} - \frac{2B}{r^3}\right)\frac{B}{r} + \frac{2B}{r^2}\frac{B}{r^2}\right]$$

$$= \frac{\mu}{\varrho}\frac{4B^2}{r^4}.$$

In the solid-body rotation of the forced vortex, no work per mass unit is done by viscous tangential stresses, and solid-body motion is the only case where this is so.[8] In the free vortex, however, work is done by the shear stresses.

FIG. 11.12

In considering the effects of shear stresses upon unit mass of fluid, the quantity μ/ϱ is seen to arise. This ratio, being of such importance, is called the kinematic coefficient of viscosity and given the symbol ν. Thus

$$\nu \equiv \mu/\varrho.$$

11.12 Turbulence

There is a certain type of unsteady viscous flow in which the velocity fluctuations are three-dimensional, small in amplitude, high in frequency, and random in nature; this type of fluctuation is called turbulence.

Fluctuations that are of a larger scale and regular are referred to as eddies.

These two characteristics are observable in the flow past a circular cylinder, which is illustrated by use of smoke as seen in Fig. 11.12. Observing the real flow, rather than just looking at this picture, it can be seen that upstream and to either side of the cylinder the flow is steady to a very high degree, but downstream vortices are shed from the cylinder at regular time intervals and alternately from top and bottom, making the flow unsteady there.

These features are indicated by plots of velocity against time. The velocity at any point can be regarded as being the sum of the time mean velocity, $\bar{q}$, and the fluctuating components, $\dot{x}'$, $\dot{y}'$, and $\dot{z}'$, in the x-, y-, and z-directions. Two diagrams, of the measured variation of the fluctuating components with time t for the flow of Fig. 11.12, are given in Fig. 11.13a, b. In each case the upper trace is of $\dot{x}'$, the fluctuating component in the general stream direction, and the lower is of $\dot{y}'$, which is in a direction perpendicular to both the stream direction and the axis of the cylinder. Figure 11.13a is of the comparatively steady flow upstream of the cylinder. Both the

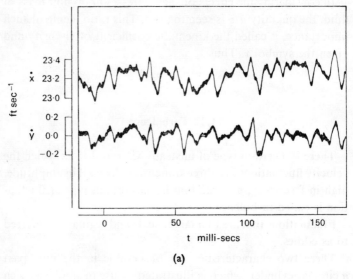

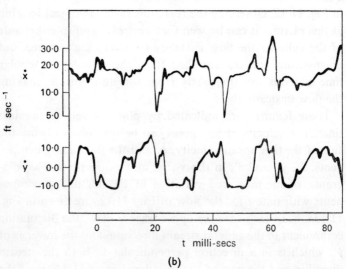

Fig. 11.13

$\dot{x}'$- and $\dot{y}'$-components are seen to be small; the measured root mean square values of the fluctuation, as a fraction of the mean velocity, were found to be

$$\frac{\sqrt{\overline{\dot{x}'^2}}}{\bar{q}} = 0{\cdot}59\%, \quad \frac{\sqrt{\overline{\dot{y}'^2}}}{\bar{q}} = 0{\cdot}33\%.$$

Figure 13b is of the unsteady eddying flow measured at the point indicated in Fig. 11.12. These show the passage of the large vortices at 20 msec intervals with the smaller scale turbulence superimposed. Now the fluctuations are much larger; the corresponding root mean square values were measured to be approximately

$$\frac{\sqrt{\overline{\dot{x}'^2}}}{\bar{q}} = 37\%, \quad \frac{\sqrt{\overline{\dot{y}'^2}}}{\bar{q}} = 49\%.$$

If a turbulent flow is then regarded as being one with fluctuations superimposed upon a distribution of mean velocity $\bar{q}$, then great care must be taken to avoid substitution of values of $\bar{q}$ into eqn. (11.10) to obtain a time average value of the shear stress $\bar{\tau}$. In general $\bar{\tau}$ is very much greater than the value that would be obtained in this way. This is because the transverse fluctuations transport momentum into adjacent portions of the fluid that are travelling at different values of $\bar{q}$. This also has the effect of reducing values of the gradient of mean velocity normal to the mean streamline, that is $\partial \bar{q}/\partial n$.

There is a notable exception to this. At a solid boundary where not only the velocity, but consequently the fluctuations, are zero, eqn. (11.11) is still valid in the form

$$\bar{\tau}_0 = \mu \left(\frac{\partial \bar{q}}{\partial n}\right)_0. \tag{11.23}$$

11.13 Boundary layer and wake flow

In a fluid flow past a slender body, such as a ship hull or an aeroplane wing, it is found that viscous rotation effects are often confined to a thin layer of the flow adjacent to the body surface. This region is called the boundary layer.

The flow past an aerofoil is sketched in Fig. 11.14. It is largely two-dimensional.

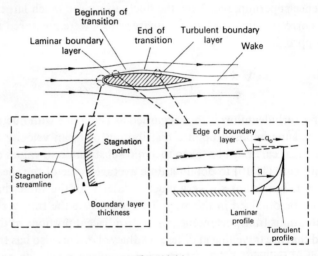

Fig. 11.14

The streamline dividing the flow between that over the upper and lower surfaces is called the stagnation streamline and it meets the nose of the aerofoil at a position called the stagnation point. This is illustrated in the inset diagram. The interesting feature of the flow along the stagnation streamline is that it behaves as if frictional rotational effects were absent; obviously, in the symmetrical flow illustrated, the rotation of an element is zero along this line, but even in a flow lacking this symmetry this frictionless behaviour is observed. In the region of the stagnation point the boundary layer has a finite thickness[9] through which the velocity rises rapidly from its zero value at the surface.†

Just downstream of the nose the boundary layer generally increases in thickness, the flow within it being steady to a high degree. Further downstream an instability appears, disturbances are amplified, and bursts of turbulent flow travel downstream increasing in frequency of occurrence in the flow direction until all the boundary layer is an unsteady turbulent flow. This is shown in Fig. 11.14 where the boundary layer thickness has been exaggerated for purposes of illustration.

The upstream region of the boundary layer where the flow is steady is called the laminar boundary layer, the downstream region of unsteady flow is called the turbulent boundary layer, and the intermediate flow is called the transition region. In other types of flow the change can be the other way, that is a turbulent viscous flow can become a laminar one.[10, 11]

There are two principal results of the change from laminar to turbulent flow in the boundary layer. The first is that the boundary layer increases in thickness and the second is that, as discussed in § 11.12, the velocity profile is flattened. The velocity profiles for laminar and turbulent flow are shown in the inset sketch in Fig. 11.14. As a result of the changed profile, the velocity gradient at the wall is much steeper for the turbulent flow and so, from eqn. (11.23), the wall shear stress is greatly increased; a factor of more than 10 is typical.

As shown in Fig. 11.14, at the downstream edge of the aerofoil the boundary layers from the upper and lower surfaces merge and travel downstream forming what is known as the wake. A physical significance of the wake is discussed in § 14.6.

Reverting again to the flow past a circular cylinder that was

† See § 11.7.

illustrated in Fig. 11.12, in this case the boundary layer persists around only part of the cylinder surface. In the region of the halfway position the boundary layer appears to leave the surface and break up into a very turbulent wake, as sketched in Fig. 11.15. This position is called the separation point.

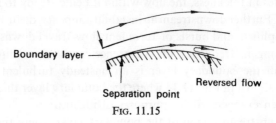

Fig. 11.15

Downstream of it the flow adjacent to the surface is reversed in direction. The physical explanation of this phenomenon is discussed further in § 14.4.

11.14 Compressible and incompressible flows

The classification into compressible or incompressible flows refers respectively to the presence or the absence of density variations in a flow pattern with change either of position or of time.

All real fluids are compressible to some degree. Only rarely in the flow of liquids has density variation to be accounted for; one exception occurs with the phenomenon of water hammer. But gases are more readily compressible. However, if all the following conditions are satisfied a gas flow can be assumed to be incompressible:

(1) The viscous stresses must be very small compared with the normal pressure stresses.[12]
(2) The application of heat must be very small.

(3) The velocity must be low. An increase of velocity from zero to about 150 ft sec^{-1} in air at atmospheric conditions is typical as resulting in a 1 per cent change in density.[6]

(4) If the flow is unsteady then pressure waves must be weak.

(5) If the flow is through a machine then the density change associated with the pressure change must be small.

References

1. HUBER, P. W. and KANTROWITZ, A., Heat capacity lag measurements in various gases, *J. Chem. Phys.* **15,** 275 (1947).
2. BINNIE, A. M. and WOODS, M. W., The pressure distribution in a convergent–divergent steam nozzle, *Proc. I. Mech. E.* **138,** 246 (1938).
3. LUKASIEWICZ, J. and ROYLE, J. K., *Effects of Air Humidity in Supersonic Wind Tunnels,* Aer. Res. Council, R. & M. 2563, June 1948.
4. LAMB, H., *Hydrodynamics,* 6th edn., Cambridge, 1932, arts. 325, 358.
5. LIGHTHILL, M. J., *Viscosity Effects in Sound Waves of Finite Amplitude,* Surveys in Mechanics (ed. G. K. Batchelor and R. M. Davies), Cambridge, 1956, p. 250.
6. VON MISES, R. V., *Mathematical Theory of Compressible Fluid Flow,* Academic Press, New York, 1958, p. 25.
7. PIERCY, N. A. V., *Aerodynamics,* Eng. Univ. Press, 1937, art. 205.
8. LAMB, H., *Hydrodynamics,* 6th edn., Cambridge, 1932, art. 329.
9. PRANDTL, L., *The Mechanics of Viscous Fluids,* Aerodynamic Theory, vol. 3, div. G (Ed. W. F. Durand), Springer, Berlin, 1935, p. 65.
10. SIBULKIN, M., *Transition from Turbulent to Laminar Pipe Flow,* Convair Scientific Res. Lab. Res. Note 52, Oct. 1961.
11. STERNBERG, J., *The Transition from a Turbulent to a Laminar Boundary Layer,* U.S. Army. Bal. Res. Lab. Aberdeen (U.S.A.) Rep. 906, May 1954.
12. TAYLOR, G. I. and SAFFMAN, P. G., Effects of compressibility at low Reynolds number, *J. Aero. Sci.* **24** (1957).

CHAPTER 12

CONSERVATION OF MASS IN A FLUID FLOW

12.1 Flow through a streamtube

Figure 12.1 is a sketch of a small element of fluid, of mass δm, that is moving along a path in a fluid flow at a velocity q. Its length in the direction of motion is δl, its cross-sectional area in a plane that is perpendicular to q is δA, and its density is ϱ. Suppose that in a time δt it moves to the adjacent position

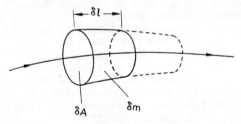

Fig. 12.1

shown dotted, so that the velocity q is the limiting value of $\delta l/\delta t$. Though it can be under a shearing action, in the limit as the size of the element tends to zero the requirements of the solid-body assumption, used in the present definition of the mass, are satisfied in that the linear motion is described by a single value of the acceleration associated with the corresponding point. This element is a system, and so its mass is conserved; the latter does not change with time as the element moves along a pathline. That is, following the element,

$$\frac{D}{Dt}(\delta m) = 0.$$

From the application of eqn. (11.5),

$$0 = \frac{D}{Dt}(\delta m) = \frac{\partial}{\partial t}(\delta m) + q\frac{\partial}{\partial l}(\delta m). \qquad (12.1)$$

The mass of the element is given by

$$\delta m = \varrho\, \delta A\, \delta l \qquad (12.2)$$

so that the second term on the right-hand side of eqn. (12.1) becomes

$$q\frac{\partial}{\partial l}(\delta m) = q\frac{\partial}{\partial l}(\varrho\, \delta A\, \delta l)$$

$$= q\, \delta l\, \frac{\partial}{\partial l}(\varrho\, \delta A) + \varrho\, \delta A\, q\, \frac{\partial}{\partial l}(\delta l).$$

Now as

$$\frac{dl}{dt} = q$$

so also

$$\frac{d}{dt}(\delta l) = \delta q,$$

a result which is interpreted by the sketch of Fig. 12.2.

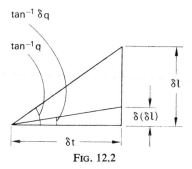

Fig. 12.2

Thus
$$q \frac{\partial}{\partial l}(\delta m) = q \; \partial(\varrho \; \delta A) + \varrho \; \delta A \; \delta q$$
$$= \partial(\varrho q \; \delta A).$$

Substitution into eqn. (12.1) then results in

$$\frac{\partial}{\partial t}(\delta m) + \partial(\varrho q \; \delta A) = 0. \tag{12.3}$$

Remembering, from its development, that the second term in this relation is a differential with respect to l at a particular instant of time, then it can be summed up for a succession of particles along a streamline at this fixed instant of time. This succession of particles will form a streamtube as illustrated in

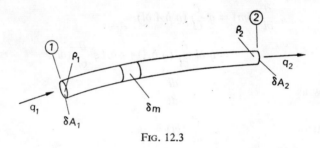

Fig. 12.3

Fig. 12.3. The first term can similarly be summed and thus integrating between two stations, say 1 and 2 as indicated in Fig. 12.3, gives

$$\int_1^2 \frac{\partial}{\partial t}(\delta m) + \int_1^2 \partial(\varrho q \; \delta A) = 0$$

or

$$\frac{\partial}{\partial t} \int_1^2 \delta m + \varrho_2 q_2 \; \delta A_2 - \varrho_1 q_1 \; \delta A_1 = 0, \tag{12.4}$$

where $\int_1^2 \delta m$ is the mass of the fluid contained within the streamtube.

The terms in eqn. (12.4) have a physical significance as follows. The first term is the rate of change of mass within the volume of the streamtube. The significance of the other two terms is illustrated by referring to Fig. 12.1. In a time δt the particles of fluid occupying a volume $\delta A\, \delta l$ have moved forward a distance δl so that a mass of fluid $\varrho\, \delta A\, \delta l$ has crossed δA. Thus the rate of mass flow $\delta \dot{m}$ is

$$\delta \dot{m} = \frac{\varrho\, \delta A\, \delta l}{\delta t} = \varrho q\, \delta A. \qquad (12.5)$$

The second term is thus the rate at which mass is flowing out of the volume across δA_2; and the third term is minus the rate at which mass is flowing in across δA. Hence eqn. (12.4) says that the net mass that flows during a time δt into a certain contour fixed in space is equal to the increase in this time of the mass contained within the contour. It gives the result that mass is additive.[†]

Equation 12.4 is a result that applies to a certain type of contour fixed in space. This contour is an example of what is called a control surface. The volume contained by this surface is called a control volume. It is seen that a relation applicable to a control volume was obtained by initially considering the characteristics of a system: such a technique will be used again in later chapters.

If the planes at the ends of a streamtube are not perpendicular to it, then eqn. (12.4) must be modified. Suppose another plane, of area δa, is drawn at an angle θ, as illustrated in Fig. 12.4. Then

$$\delta A = \delta a \cos \theta.$$

[†] Then, using eqn. (1.8), it can be shown that forces are additive.

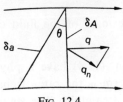

Fig. 12.4

But also

$$q_n = q \cos \theta,$$

where q_n is the component of the velocity in a direction perpendicular to the oblique plane. Thus

$$q \, \delta A = q_n \, \delta a,$$

and so eqns. (12.5) and (12.4) can be rewritten

$$\delta \dot{m} = \varrho q_n \, \delta a, \qquad (12.6)$$

and

$$\frac{\partial}{\partial t} \int_1^2 \delta m + \varrho_2 q_{n2} \, \delta a_2 - \varrho_1 q_{n1} \, \delta a_1 = 0. \qquad (12.7)$$

Writing the volume of the small element δV, then

$$\delta m = \varrho \, \delta V,$$

and so eqn. (12.7) can be rewritten

$$\frac{\partial}{\partial t} \int_1^2 \varrho \, dV + \varrho_2 q_{n2} \, \delta a_2 - \varrho_1 q_{n1} \, \delta a_1 = 0. \qquad (12.8)$$

Equation 12.3 can be simplified when the flow is along a straight streamtube of constant cross-sectional area in, say, the x-direction. Then if the element is of length δx and the

CONSERVATION OF MASS IN A FLUID FLOW   209

velocity is now $\dot{x}$,
$$\delta m = \varrho \, \delta A \, \delta x$$
and
$$\partial(\varrho \dot{x} \, \delta A) = \delta A \frac{\partial}{\partial x}(\varrho \dot{x}) \, \delta x.$$

Equation 12.3 then becomes

$$\frac{\partial \varrho}{\partial t} + \frac{\partial}{\partial x}(\varrho \dot{x}) = 0. \tag{12.9}$$

12.2 Flow through a control volume

A control volume is a volume drawn within a fluid flow and whose surface is fixed in space so that fluid crosses at least part of it. One such is illustrated in Fig. 12.5 where some of the streamlines of the flow through it are shown. Certain

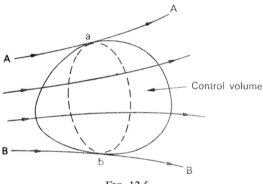

Fig. 12.5

streamlines, such as those marked *AA* and *BB* will be tangential to the surface of the control volume, and all the tangential points of contact, such as the two marked *a* and *b*, will form a complete circuit as shown dotted in this figure. To the left of this circuit fluid will be flowing into the control volume, to the right it will be flowing out.

The complete flow is built up of streamtubes cut off at each end by the surface of the control volume, and so eqn. (12.8) can be applied to this flow. It can be written down separately for each streamtube of the flow through the control volume and then all these equations can be added up. Summation of all the first terms of the eqns. (12.8) gives the rate of change of the total mass within the control volume. This can be written,

$$\frac{\partial}{\partial t} \int \varrho \, dV,$$

where now the summation is over the whole control volume.

Summation of all the second terms of eqn. (12.8) will give the sum of all the individual rates of mass flow out of each streamtube and so will give the total rate of mass flow out of the control volume. This is written

$$\int_{\text{out}} \varrho q_n \, da.$$

Similarly, summation of all the final terms of eqn. (12.8) will, ignoring the negative sign, give the total rate of mass flow into the control volume which is written

$$\int_{\text{in}} \varrho q_n \, da.$$

So the continuity equation for a control volume can be written

$$\frac{\partial}{\partial t} \int \varrho \, dV + \int_{\text{out}} \varrho q_n \, da - \int_{\text{in}} \varrho q_n \, da = 0. \qquad (12.10)$$

12.3 Zero time derivative

The first term in eqn. (12.10) is zero in the following two cases:

(a) If the flow is compressible but steady. Then the density ϱ at any point within the control volume is a constant with change in time and so the contained mass stays a constant.
(b) If the flow is incompressible, being either steady or unsteady, then the density varies neither with position nor with time and so the contained mass is constant.

In case (a) eqn. (12.10) becomes

$$\int_{\text{out}} \varrho q_n \, da = \int_{\text{in}} \varrho q_n \, da \qquad (12.11)$$

so that these mass flow rates are equal.

In case (b), eqn. (12.11) simplifies further, because ϱ is a constant, to

$$\int_{\text{out}} q_n \, da = \int_{\text{in}} q_n \, da. \qquad (12.12)$$

These terms are ones of rate of volume flow.

12.4 Mean velocity

A mean velocity taken over a plane across the flow through a duct can aid simple calculation. Such a plane is illustrated in Fig. 12.6.

When the flow is compressible such a mean can be defined by

$$\bar{q} \equiv \frac{1}{\bar{\varrho} a} \int \varrho q_n \, da.$$

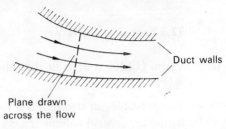

Fig. 12.6

This involves the need for a definition of a mean density which is considered again in § 16.7. Also $\bar{q}$ is dependent, through the value of a, upon the choice of the plane of integration.

When the flow is incompressible this definition is simplified to

$$\bar{q} \equiv \frac{1}{a} \int q_n \, da.$$

12.5 Application of the continuity equation

The distinction between the concept of a system and that of a control volume is important. This is now illustrated by investigating the expansion—by a piston—of some gas contained in a cylinder. This is illustrated in Fig. 12.7.

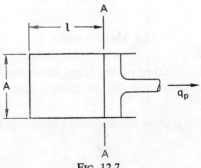

Fig. 12.7

If the gas in the cylinder is regarded as a system, then its mass is a constant or

$$\varrho V = \text{constant}.$$

If the piston is moving to the right with a velocity q_p that is sufficiently low to allow ϱ to be considered a constant throughout the system, then differentiating the previous relation,

$$\varrho \frac{dV}{dt} + V \frac{d\varrho}{dt} = 0. \tag{12.13}$$

Also

$$V = Al,$$

where A is the cross-sectional area of the piston and l is the enclosed length of the cylinder. Differentiating this gives

$$\frac{dV}{dt} = A \frac{dl}{dt}$$
$$= A q_p. \tag{12.14}$$

So that, eliminating dV/dt between eqns. (12.13) and (12.14),

$$\frac{d\varrho}{dt} = -\frac{\varrho}{V} A q_p. \tag{12.15}$$

In contrast, at any instant the volume of the gas can be regarded as a control volume, so that eqn. (12.10) can be used. Noting that the volume of a control volume is fixed, and retaining the assumption that ϱ is uniform throughout the control volume, then the first term of eqn. (12.10) is

$$\frac{\partial}{\partial t}(\varrho V) = V \frac{\partial \varrho}{\partial t}.$$

There is no flow into the control volume so that

$$\int_{\text{in}} \varrho q_n \, da = 0.$$

There is a flow out of the control volume because the gas at that part of the control surface formed by the piston face, denoted AA in Fig. 12.7, is moving to the right with a velocity q_p. Thus

$$\int_{\text{out}} \varrho q_n \, da = \varrho q_p A.$$

Thus eqn. (12.10) gives

$$V \frac{d\varrho}{dt} = -\varrho q_p A \quad \text{or} \quad \frac{d\varrho}{dt} = -\frac{\varrho}{V} A q_p.$$

Comparison of this result with that of eqn. (12.15) shows how the system and control volume concepts give the same result.

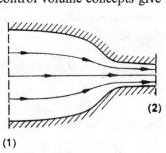

Fig. 12.8

Another simple application of the continuity equation can be made to the incompressible flow through a rapidly contracting duct along the walls of which there is a thin boundary layer. A sketch of this is given in Fig. 12.8. Across the upstream and downstream sections, denoted (1) and (2) in this figure, the flow is closely uniform so that drawing a control volume coincident with the contraction walls and the planes (1) and (2) enables eqn. (12.11) to be written as

$$\int_2 \varrho q_2 \, da = \int_1 \varrho q_1 \, da$$

and thus

$$\bar{q}_2 a_2 = \bar{q}_1 a_1.$$

CONSERVATION OF MASS IN A FLUID FLOW

It is worth noting that the approximation of this last equation requires a closely uniform flow only at stations (1) and (2); in between these positions the flow is not so restricted.

12.6 Diffusion

If two quantities of different fluids—either two liquids or two gases—are placed in contact, and if their pressures and temperatures are equal so that there is mechanical and thermal equilibrium between them, then these two fluids will, of their own accord, intermix on a molecular scale. This process is known as diffusion.

The intermingling of the two fluids prohibits the use of the concept of a system as defined in § 5.1. Consequently, the proof of the continuity equation given in § 12.1 is not valid.

At any time after intermingling has commenced, an element of the mixture will have, in the notation of § 9.4, N molecules in a volume of V. The concentration C is defined by

$$C \equiv \frac{N}{V}. \qquad (12.16)$$

For two fluids, denoted 1 and 2, this concentration can, as indicated in § 9.4, be written

$$C = \frac{N_1}{V} + \frac{N_2}{V}.$$

Then, defining the concentration of the constituents C_1 and C_2 by

$$C_1 \equiv \frac{N_1}{V} \quad \text{and} \quad C_2 \equiv \frac{N_2}{V}, \qquad (12.17)$$

there results

$$C = C_1 + C_2 \qquad (12.18)$$

The mass of a quantity of fluid is $NA_m M$, which equals $CVA_m M$. The number of moles is thus CVA_m, so that the number of moles per unit volume is CA_m. This is called the molar concentration c and so

$$c = CA_m. \tag{12.19}$$

This applies to the constituents of a mixture, so that, from eqn. (12.18),

$$c = c_1 + c_2. \tag{12.20}$$

The density ϱ is given by

$$\varrho = \frac{CVA_m M}{V}$$
$$= cM, \tag{12.21}$$

so that

$$\varrho = \varrho_1 + \varrho_2$$
$$= c_1 M_1 + c_2 M_2, \tag{12.22}$$

where the partial densities ϱ_1 and ϱ_2 are each based upon the full volume V.

During diffusion, the molecules of the two constituents have mean drift velocities v_1 and v_2. This is illustrated in Fig. 12.9 for the simple case of diffusion in only the x-direction

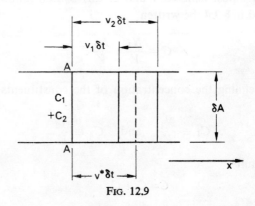

Fig. 12.9

along a streamtube of cross-section δA. After a time δt, the molecules that occupied the section denoted AA have moved average distances of $v_1\,\delta t$ and $v_2\,\delta t$. The corresponding volumes occupied by succeeding molecules are $v_1\,\delta t\,\delta A$ and $v_2\,\delta t\,\delta A$. The number of molecules N that have moved across AA is thus

$$\begin{aligned}N &= N_1 + N_2 \\ &= (c_1 v_1 + c_2 v_2)\frac{\delta A\,\delta t}{A_m}.\end{aligned} \qquad (12.23)$$

This total number can be expressed by

$$N = cv^*\frac{\delta A\,\delta t}{A_m},$$

which defines the mean concentration velocity v^*. Noting eqn. (12.20), it follows that

$$v^* = \frac{c_1 v_1 + c_2 v_2}{c_1 + c_2}. \qquad (12.24)$$

The total rate per unit area, at which molecules are crossing AA, is $N/(\delta A\,\delta t)$. When this is zero, $v^* = 0$, and so from eqn. (12.24)

$$c_1 v_1 + c_2 v_2 = 0. \qquad (12.25)$$

Diffusion is described as the motion of particles at a velocity that is relative to v^*. That is, it is a motion relative to a condition of zero total flow of particles. A diffusion rate for the ith constituent $\dot{n}_i$ is defined by

$$\dot{n}_i \equiv c_i(v_i - v^*). \qquad (12.26)$$

The sum of these rates is, by using eqn. (12.24), given by

$$\dot{n}_1 + \dot{n}_2 = c_1(v_1 - v^*) + c_2(v_2 - v^*) = 0, \qquad (12.27)$$

which repeats the statement just made that diffusion is relative to zero total particle flow; the particle diffusion rates of the two constituents are equal and opposite. Also this further amplifies the previous statement that diffusion prohibits the use of the concept of a system.

12.7 Conservation of numbers of particles

Figure 12.10 is another sketch of a small streamtube, of cross-sectional area δA, along which flow and diffusion are taking place in only the x-direction.

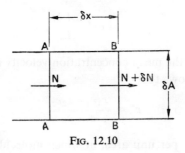

Fig. 12.10

In time δt the number of particles N moving across section AA into the element is given by eqn. (12.23). If the length of the element is δx then, correspondingly, the number moving out across section BB is

$$N + \delta N$$
$$= N + \frac{\partial N}{\partial x} \delta x$$
$$= \left[(c_1 v_1 + c_2 v_2) + \frac{\partial}{\partial x} (c_1 v_1 + c_2 v_2) \delta x \right] \frac{\delta t \, \delta A}{A_m}.$$

Hence the net number moving in is

$$-\frac{\partial}{\partial x} (c_1 v_1 + c_2 v_2) \frac{\delta t \, \delta A \, \delta x}{A_m}.$$

CONSERVATION OF MASS IN A FLUID FLOW

The number of particles within the element at any time is $C \, \delta A \, \delta x$, and so, noting eqns. (12.15) and (12.16), the change in time δt is

$$\frac{\partial}{\partial t}(C_1+C_2) \, \delta A \, \delta x \, \delta t$$

$$= \frac{\delta A \, \delta x \, \delta t}{A_m} \frac{\partial}{\partial t}(c_1+c_2).$$

The net number of particles moving into the element must equal the change within, and thus

$$-\frac{\partial}{\partial x}(c_1 v_1 + c_2 v_2) = \frac{\partial}{\partial t}(c_1+c_2).$$

Substitution from eqns. (12.16) and (12.19) gives

$$\frac{\partial c}{\partial t} + \frac{\partial}{\partial x}(cv^*) = 0. \tag{12.28}$$

A simple solution to this equation is obtainable for the case of diffusion between two gases whose temperature and total pressure remain constant. Then from eqns. (9.17) and (12.15), $c = $ constant, and so eqn. (12.28) reduces to

$$\frac{\partial v^*}{\partial x} = 0 \quad \text{or} \quad v^* = \text{constant}.$$

If there is a solid wall at some value of x, then $N = 0$ there, and so, from eqns. (12.23) and (12.24), $v^* = 0$ at the wall and hence, from above, $v^* = 0$ for all values of x.

12.8 Conservation of mass

The momentum of a particle of the ith constituent due to its drift velocity is $M_i A_m v_i$. The sum of the momenta for N_i particles is thus

$$M_i A_m v_i N_i$$
$$= M_i A_m v_i C_i V$$
$$= M_i c_i v_i V$$
$$= \varrho_i v_i V,$$

where use has been made of eqns. (12.17), (12.19), and (12.21). Thus the total momentum per unit volume for both constituents is

$$\varrho_1 v_1 + \varrho_2 v_2.$$

This can be expressed by $\varrho \dot{x}$ thus defining this momentum mean velocity $\dot{x}$ by

$$\varrho \dot{x} \equiv \varrho_1 v_1 + \varrho_2 v_2,$$

which, using eqn. (12.22), can be written

$$\dot{x} = \frac{\varrho_1 v_1 + \varrho_2 v_2}{\varrho_1 + \varrho_2} \tag{12.29}$$

or
$$\dot{x} = \frac{M_1 c_1 v_1 + M_2 c_2 v_2}{M_1 c_1 + M_2 c_2}. \tag{12.30}$$

Considering again Fig. 12.10, the mass rate at which the ith component is flowing across AA is, from eqns. (12.6) and (12.21), $M_i c_i v_i \, \delta A$. The total rate is thus

$$(M_1 c_1 v_1 + M_2 c_2 v_2) \, \delta A,$$

and so, by analogy with the discussion of § 12.7, the net mass into the element in time δt is

$$-\frac{\partial}{\partial x}(M_1 c_1 v_1 + M_2 c_2 v_2) \, \delta x \, \delta A \, \delta t.$$

CONSERVATION OF MASS IN A FLUID FLOW

The mass within the element is $(M_1c_1 + M_2c_2)\,\delta A\,\delta x$, and this changes in time δt by the amount

$$\frac{\partial}{\partial t}(M_1c_1 + M_2c_2)\,\delta t\,\delta A\,\delta x.$$

Mass being additive, these two terms must sum to zero so that

$$\frac{\partial}{\partial t}(M_1c_1 + M_2c_2) + \frac{\partial}{\partial x}(M_1c_1v_1 + M_2c_2v_2) = 0, \quad (12.31)$$

which, with substitution from eqns. (12.22) and (12.30), becomes

$$\frac{\partial \varrho}{\partial t} + \frac{\partial}{\partial x}(\varrho \dot{x}) = 0.$$

This is seen to be identical in form to eqn. (12.9), but now $\dot{x}$ is defined by eqn. (12.30); and now this is not the mass convection velocity. This can be seen as follows. The total mass flow rate is

$$(M_1c_1v_1 + M_2c_2v_2)\,\delta A.$$

The diffusion mass flow rate is

$$[M_1c_1(v_1 - v^*) + M_2c_2(v_2 - v^*)]\,\delta A.$$

The difference between these two quantities is the convection flow rate which is then

$$(M_1c_1 + M_2c_2)v^*\,\delta A$$
$$= \varrho v^*\,\delta A,$$

so that v^* is the mass flow convection velocity.

Thus, although the equation reduces to the form of the standard continuity equation, it is important to remember the result of § 12.7 that it is not $\dot{x}$ but v^* that is zero at a solid boundary.

Considering again the case when c is constant, and using eqns. (12.20) and (12.24), eqn. (12.31) becomes

$$\frac{\partial}{\partial t}[(M_1-M_2)c_1+M_2c]+\frac{\partial}{\partial x}[(M_1-M_2)c_1v_1+M_2cv^*] = 0,$$

and thus, as M_1, M_2, c, and v^* are all constants, this becomes

$$\frac{\partial c_1}{\partial t}+\frac{\partial}{\partial x}(c_1v_1) = 0. \qquad (12.32)$$

12.9 Coefficients of diffusion

The diffusion rate $\dot{n}$ as defined in eqn. (12.26) is found experimentally to be related to the gradient of the concentration by a relation of the form

$$\dot{n}_1 = -cD_{12}\frac{\partial}{\partial x}\left(\frac{c_1}{c}\right) \qquad (12.33)$$

for the 1 constituent diffusing into the 2 constituent in the x-direction, and for the other direction by

$$\dot{n}_2 = -cD_{21}\frac{\partial}{\partial x}\left(\frac{c_2}{c}\right). \qquad (12.34)$$

The terms D_{12} and D_{21} are known as the diffusion coefficients. They are properties, being, for gases, mainly a function of the pressure and the temperature and, for liquids, mainly of the proportion of the constituent to the mixture.

Diffusion can also occur as a result of either a pressure gradient or a temperature gradient.[1] The discussion of this chapter is limited to the case of diffusion resulting from a gradient of concentration.

CONSERVATION OF MASS IN A FLUID FLOW

Substitution of eqns. (12.33) and (12.34) into (12.27) gives

$$-c\left[D_{12}\frac{\partial}{\partial x}\left(\frac{c_1}{c}\right)+D_{21}\frac{\partial}{\partial x}\left(\frac{c_2}{c}\right)\right]=0$$

and, using eqn. (12.20),

$$D_{12}\frac{\partial}{\partial x}\left(\frac{c_1}{c}\right)+D_{21}\frac{\partial}{\partial x}\left(1-\frac{c_1}{c}\right)=0,$$

so that

$$D_{12}\frac{\partial}{\partial x}\left(\frac{c_1}{c}\right)-D_{21}\frac{\partial}{\partial x}\left(\frac{c_1}{c}\right)=0$$

and then
$$D_{12} = D_{21}. \qquad (12.35)$$

Differentiating eqn. (12.26) and using eqn. (12.33) gives

$$\frac{\partial}{\partial x}(c_1 v_1) = \frac{\partial}{\partial x}(\dot{n}_1 + c_1 v^*)$$

$$= \frac{\partial}{\partial x}\left[-cD_{12}\frac{\partial}{\partial x}\left(\frac{c_1}{c}\right)+c_1 v^*\right].$$

Again, considering the case $c = $ constant so that also $v^* = $ constant, substitution of this expression into eqn. (12.32) result in

$$\frac{\partial c_1}{\partial t} = \frac{\partial}{\partial x}\left(D_{12}\frac{\partial c_1}{\partial x}\right) - v^*\frac{\partial c_1}{\partial x}.$$

If D_{12} can be taken as having a constant value[†] and if the origin of x is chosen to make $v^* = 0$, then, finally,

$$\frac{\partial c_1}{\partial t} = D_{12}\frac{\partial^2 c_1}{\partial x^2}.$$

A simple solution to this equation is obtained for steady-state

---

[†] Cases when D_{12} cannot be taken as a constant are considered in ref. 2. In certain cases use of a mean value of D gives exact results.[3]

diffusion for which $\partial c_1/\partial t = 0$. Then, a double integration of the right-hand side gives

$$D_{12}c_1 = \alpha + \beta x,$$

where α and β are constants of integration. In this case the concentrations vary linearly.

12.10 The semi-permeable membrane

Often in chemical processes a separation of two substances in a mixture is obtained by use of a membrane that permits the passage of, say, constituent 1 whilst preventing the passage of constituent 2. Consequently, at the membrane, $v_2 = 0$ and so $\dot{n}_2 = -c_2 v^*$. The mass flow rate of constituent 1 is

$$\begin{aligned}
M_1 & c_1 v_1 \, \delta A \\
&= M_1 c_1 \, \delta A (v_1 - v^* + v^*) \\
&= M_1 c_1 \, \delta A \left(\frac{\dot{n}_1}{c_1} - \frac{\dot{n}_2}{c_2} \right) \\
&= M \, \delta A \left[-c D_{12} \frac{\partial}{\partial x} \left(\frac{c_1}{c} \right) + \frac{c_1}{c_2} c D_{21} \frac{\partial}{\partial x} \left(\frac{c_2}{c} \right) \right] \\
&= M_1 \, \delta A \, c D_{12} \left[\frac{c_1}{c_2} \frac{\partial}{\partial x} \left(\frac{c_2}{c} \right) - \frac{\partial}{\partial x} \left(\frac{c_1}{c} \right) \right] \\
&= M_1 \, \delta A \, c D_{12} \left[\frac{c_1}{c_2} \frac{\partial}{\partial x} \left(1 - \frac{c_1}{c} \right) - \frac{\partial}{\partial x} \left(\frac{c_1}{c} \right) \right] \\
&= -M_1 \, \delta A \, c D_{12} \left(1 + \frac{c_1}{c_2} \right) \frac{\partial}{\partial x} \left(\frac{c_1}{c} \right) \\
&= -M_1 \, \delta A \, c D_{12} \frac{1}{1 - \frac{c_1}{c}} \frac{\partial}{\partial x} \left(\frac{c_1}{c} \right) \\
&= M_1 \, \delta A \, c D_{12} \frac{\partial}{\partial x} \left[\log \left(1 - \frac{c_1}{c} \right) \right],
\end{aligned}$$

where the differential coefficient is evaluated at the membrane.

12.11 Self-diffusion

In a gas composed of a single constituent, a gradient of density results in a gradient of concentration and, correspondingly, the gas diffuses in itself. The total rate of mass flow is given by

$$Mcv \, \delta A$$
$$= M[cv^* + \dot{n}] \, \delta A.$$

For this self-diffusion, eqn. (12.33) becomes

$$\dot{n} = -D \frac{\partial c}{\partial x}, \qquad (12.36)$$

so that the total rate of mass flow is

$$M \left[cv^* - D \frac{\partial c}{\partial x} \right] \delta A$$
$$= \varrho v^* \, \delta A - D \frac{\partial \varrho}{\partial x} \, \delta A.$$

As shown in § 12.8, the first term is the convection term.

For self-diffusion in gases, $D\varrho/\mu \simeq 1\cdot 4$,[4] so that for air $D = 0\cdot 22 \times 10^{-3}$ ft^2 sec^{-1}. If the second term is to be negligible by a factor of 100, then

$$\varrho v^* = 100 D \frac{\partial \varrho}{\partial x}.$$

With a typical velocity of 10^3 ft sec^{-1} in a flow of air in which density changes are appreciable, then

$$\frac{1}{\varrho} \frac{\partial \varrho}{\partial x} = \frac{10^3}{10^2 \times 0\cdot 22 \times 10^{-3}} = 4\cdot 5 \times 10^4 \text{ ft}^{-1} = 375 \text{ in}^{-1},$$

a criterion which is less severe by a factor of about 10 than the continuum criterion described in § 2.2. Thus, usually, eqn. (12.4) is valid for the flow of a gas containing only one constituent.

References

1. BIRD, R. B., STEWART, W. E., and LIGHTFOOT, E. N., *Transport Phenomena*, Wiley, New York, 1960, p. 564.
2. CRANK, J., *The Mathematics of Diffusion*, Oxford, 1956, p. 147.
3. TRIVEDI, R. and POUND, G. M., Effect of concentration-dependent diffusion coefficient on the migration of interphase boundaries, *J. Appl. Phys.* **38** (9), 3569 (Aug. 1967).
4. CHAPMAN, S. and COWLING, T. G., *The Mathematical Theory of Non-uniform Gases*, Cambridge, 1960, p. 251.

CHAPTER 13

THE EQUATIONS RELATING PROCESS PHENOMENA

13.1 Summary of the basic equations

An equation that relates two features of a process exists for each of several phenomena. Several such equations have been presented here and they are as follows.

First, eqn. (1.8) relates the force acting on a rigid system to the acceleration of that system and is

$$F = m \frac{d}{dx}\left(\frac{1}{2} q^2\right). \qquad (13.1)\,[1.8]$$

Second, eqn. (8.2) relates the normal stress in an elastic solid to the strain and is

$$\sigma = E\varepsilon, \qquad (13.2)\,[8.2]$$

whilst the tangential stress is related to the shear strain by

$$\tau = N \frac{dx}{dy}. \qquad (13.3)$$

Fourth, in a Newtonian fluid, eqn. (5.1) relates the tangential stress to the velocity gradient and is

$$\tau = \mu \frac{dq}{dy}. \qquad (13.4)\,[5.1]$$

Fifth, eqn. (7.1) relates the rate of heat intensity to the temperature gradient and is

$$\eta = -k \frac{dT}{dx}. \qquad (13.5)\,[7.1]$$

Sixth, the electric current intensity j is related to the voltage gradient $d\varphi/dx$ by Ohm's law, which is

$$j = \lambda \frac{d\varphi}{dx}, \qquad (13.6)$$

where λ is the conductivity.

And seventh, eqn. (12.33) relates the diffusion transport rate to the concentration gradient and is, for constant c,

$$\dot{n}_i = -D \frac{dc_i}{dx}. \qquad (13.7)\,[12.33]$$

These seven relations are closely analogous. In each case they give a relation between two quantities that are associated with the nature of a process. Also this relation is linear in form and is a vector relation in that the directions of these two quantities are the same.[†] Further, the coefficients relating these two quantities, which are m, E, N, μ, k, λ, and D, are each independent of the nature of the process, being purely a property. Thus, in each case the ratio of two quantities, both of which are a measure of the form of the process, is independent of that form. For example, in Ohm's law, eqn. (13.6), both the current intensity j and the voltage gradient $d\varphi/dx$ are associated with the electrical nature of a process and only exist during the process; but their ratio λ, the conductivity, is independent of the electrical phenomena, being just a property.[‡]

[†] It has already been pointed out that in the case of heat the phenomena must occur in isolation for this identity of direction to occur (§ 7.6). A similar restriction must be made to application of eqn. (13.6) in that a transverse magnetic field must not be present.

[‡] In the case of λ it is necessary to include the charge density as a property defining the state.[1]

Amongst these properties, the mass has a special feature because, for any system, it cannot be changed. Similarly, in the study of electricity, the total electrical charge contained by a system cannot be changed.

Because of the marked similarity between the preceding equations, the mathematical study of the different phenomena that they govern can often be identical. This applies particularly to the phenomena governed by eqns. (13.2) and (13.3), on the one hand, and to eqn. (13.4), on the other, and again to the phenomena governed by eqns. (13.5), (13.6), and 13.7.

13.2 Interaction between phenomena

The previous relations, eqns. (13.1) to (13.7), do not necessarily apply when more than one of the phenomena that they describe are acting simultaneously. For example, if there is both a temperature gradient and a voltage gradient in an electrical conductor, then the electrical current intensity rate at any section of the conductor is given by a relation of the form

$$j = A_1 \frac{d\varphi}{dx} + A_2 \frac{dT}{dx}. \tag{13.8}$$

The first term on the right-hand side of this relation is a measure of the drift of charged particles that results from the electric field force acting upon them, which latter is measured by the voltage gradient $d\varphi/dx$. The second term on the right-hand side is a measure of the drift of charged particles that results from the diffusion that occurs in the presence of the temperature gradient dT/dx.[†] The existence of these two drift components means that the section of the conductor being considered forms part of the boundary of a control volume and not part of the boundary of a system.

[†] See § 12.9.

As well as carrying charge, the particles that drift along the conductor carry internal energy. This rate of transport of the internal energy $\dot{e}_d$ can be written in the form

$$\dot{e}_d = B_1 \frac{dT}{dx} + B_2 \frac{d\varphi}{dx}. \qquad (13.9)$$

It must be noted that the first term in this equation is not that due to heat by thermal conduction. Heat, as defined in § 5.12, is not a convection phenomenon.

The coefficients A_1 and B_1 are those that refer to the two phenomena of electrical current flow and thermal diffusion when these exist in isolation. In certain cases coefficients like A_2 and B_2 are related by a theorem due to Onsager.[2] †

Relations like eqn. (13.8) form the basis of the study of thermoelectric effects.

References

1. GIBBINGS, J. C. and HIGNETT, E. T., Dimensional analysis of electrostatic streaming current, *Electrochimica Acta.* **2**, 819 (1966).
2. ONSAGER, L., Reciprocal relations in irreversible processes, I, *Phys. Rev.* **37** (4), 405 (15 Feb. 1931); II, **38** (12), 2265 (15 Dec. 1931).
3. LEE, J. F. and SEARS, F. W., *Thermodynamics*, Addison-Wesley, Reading, Massachusetts, 2nd edn., 1962, p. 212.
4. BIRD, R. B., STEWART, W. E. and LIGHTFOOT, E. M., *Transport Phenomena*, Wiley, New York, 1960, para. 18.4, p. 563.

---

† In analogy with eqn. (13.9) other properties can be convected. Two such are enthalpy (§ 15.8) and entropy, which latter is introduced in Montgomery, *op. cit.* Convection of electric charge and entropy is an example where the coefficients A_2 and B_2 can be related by the Onsager theorem. An expression for heat is excluded from such considerations.[3] Further reading is left to ref. 4 and the other references cited.

CHAPTER 14

THE MOMENTUM EQUATION FOR FLUIDS IN MOTION

14.1 Flow of an element of fluid

Figure 14.1 is a sketch of an element of fluid, of volume δV, that is moving with a velocity q within a flow. The length of the element is denoted δl with a corresponding vertical component δz, and its weight is δF_g acting at an angle α to the pathline.

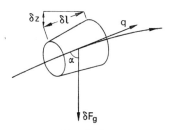

Fig. 14.1

Applying eqn. (9.33) gives the load on the element, in the direction of motion and due to the pressure distribution, as

$$-\frac{\partial p}{\partial l}\,\delta V.$$

The mass of the element is $\varrho\,\delta V$, its weight is $\varrho g\,\delta V$, and the component of this force in the direction of motion is

$$-\varrho g\,\delta V \cos\alpha.$$

If attention is limited to the type of flow discussed in § 11.11, where there is no net force on the element due to shear stresses, then application of eqn. (1.8) to the motion of this element along its path leads to

$$\frac{\partial p}{\partial l} \delta V + \varrho g\, \delta V \cos \alpha = \varrho\, \delta V\, \frac{Dq}{Dt}.$$

Noting that

$$\frac{\delta z}{\delta l} = \cos \alpha,$$

then this relation becomes

$$\frac{1}{\varrho} \frac{\partial p}{\partial l} + g \frac{dz}{dl} + \frac{Dq}{Dt} = 0, \qquad (14.1)$$

where the first and second terms represent a force per unit mass.

Further limitation of the discussion to flows that are steady has two consequences: first, as discussed in § 11.3, the pathline becomes a streamline, and second, from eqn. (11.6), the acceleration becomes

$$\frac{Dq}{Dt} = q \frac{\partial q}{\partial l}.$$

Thus eqn. (14.1) becomes

$$\frac{1}{\varrho} \frac{\partial p}{\partial l} + g \frac{dz}{dl} + q \frac{\partial q}{\partial l} = 0. \qquad (14.2)$$

This equation relates the gradients of velocity and of pressure.

Integration of eqn. (14.2) must be done with respect to l, that is along a streamline. Doing this results in

$$\int \frac{1}{\varrho} \frac{\partial p}{\partial l} dl + gz + \frac{1}{2} q^2 = \text{constant}, \qquad (14.3)$$

where the constant is so along a streamline. It can vary between streamlines.

The first integral in eqn. (14.3) can only be evaluated if the process that the element undergoes as it travels along is known in sufficient detail so that ϱ can be expressed as a single-valued function of p. The simplest case is that of incompressible flow for which $\varrho = $ constant and then eqn. (14.3) becomes

$$\frac{p}{\varrho} + gz + \frac{1}{2} q^2 = \text{constant.} \quad (14.4)$$

This result, known as the Bernoulli equation, is derived by following an element along a streamline in steady flow. Thus, as explained in § 11.3, it provides a relation applicable to all elements at any time along the streamline.

Commonly, the constant in eqn. (14.4) is expressed in two ways. It is written

$$\frac{p}{\varrho} + gz + \frac{1}{2} q^2 = \frac{P}{\varrho}$$

and P is called the total pressure. Or it is written

$$\frac{p}{\varrho} + gz + \frac{1}{2} q^2 = gZ_H,$$

and Z_H is called the total head.

Remembering that the right-hand side of eqn. (14.4) is constant and that g was assumed constant along the streamline,[†] the height z can be measured above any arbitrary datum. The physical significance of the total pressure is that it is the value of the pressure corresponding to zero velocity and zero height; if there is a stagnation point within the flow and if it is at the chosen datum level, then the pressure there is P. In liquid flows it is often convenient to express pressures in terms of heights of liquids; the total head is then the height corresponding to the total pressure.

[†] Even in meteorological studies this can usually be an acceptable approximation.

14.2 Relative values of the terms in the Bernoulli equation

Typical values of the three terms on the left-hand side of eqn. 14.4 are given in Table 14.1

TABLE 14.1

Atmospheric altitude:	0	10^4	10^5	ft
Air (p/ϱ)	0.89×10^6	0.75×10^6	0.68×10^6	ft^2 sec^{-2}
Water (p/ϱ)	1.1×10^3	—	—	ft^2 sec^{-2}
z	1	10	10^2	ft
gz	32	3.2×10^2	3.2×10^3	ft^2 sec^{-2}
q	1	10	10^2	ft sec^{-1}
$\tfrac{1}{2} q^2$	0.5	50	5×10^3	ft^2 sec^{-2}

In Table 14.1 atmospheric pressure has been taken as a typical pressure; in the case of water flow the value at zero altitude has been used.

Several features that are typical of many flows of gases and of liquids are revealed by this tabulation. In an incompressible gas flow, with the speed limitation discussed in § 11.14, the term in p/ϱ is the dominant one. Thus pressure changes are very small compared with absolute pressures, and so pressure gradients, which result in the force upon an element, are small compared with pressures. This explains why shear stresses often cannot be ignored even though they are usually very small compared with pressures, a point discussed further in § 14.6. In a gas flow the height changes are often small and the velocities fairly high, so that then the gz term is negligible compared with the

THE MOMENTUM EQUATION FOR FLUIDS IN MOTION 235

$\frac{1}{2} q^2$ term. In this case the Bernoulli equation simplifies to

$$\frac{p}{\varrho} + \frac{1}{2} q^2 = \frac{P}{\varrho}.$$

In liquid flows the three terms are often comparable and, in contrast, pressure changes can be large compared with absolute pressures. Sometimes changes in height are large and the velocities are low so that the $\frac{1}{2} q^2$ term can be neglected and then the relation

$$\frac{p}{\varrho} + gz = gZ_H$$

becomes an acceptable approximation.

14.3 Variation of properties normal to streamlines

In the discussion of § 11.11, on the flow in a free vortex, it was shown that the force per unit mass exerted by the shear stresses and in the streamline direction was zero. Equally, by inspection again of Fig. 11.11, (p. 193) it can be seen that the shear forces on the sides AB and DC are equal and opposite so that there is no net shear force in a direction perpendicular to the streamline. Referring to Fig. 14.2, which is a sketch of an element of fluid travelling along a streamline, and considering forces normal to the streamline, the pressure force outwards is

$$-\frac{\partial p}{\partial n} \delta V$$

and the weight force δF_g is

$$-\varrho g \, \delta V \sin \alpha.$$

The acceleration in this direction being $-q^2/r$, then, as before,

$$-\frac{\partial p}{\partial n} \delta V - \varrho g \, \delta V \sin \alpha = -\frac{q^2}{r} \varrho \, \delta V. \qquad (14.5)$$

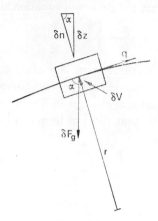

Fig. 14.2

But, as can be seen in Fig. 14.2,

$$\sin \alpha = \frac{\delta z}{\delta n},$$

and thus

$$\frac{\partial p}{\partial n} + \varrho g \frac{dz}{dn} - \frac{\varrho q^2}{r} = 0. \quad (14.6)$$

This provides a relation for the pressure gradient normal to the streamlines.

The radius of curvature r can be eliminated from eqn. (14.6) by use of eqn. (11.15), giving

$$2q\omega = \frac{1}{\varrho}\frac{\partial p}{\partial n} + g\frac{dz}{dn} + q\frac{\partial q}{\partial n}$$

which for ϱ and g constant becomes

$$2q\omega = \frac{\partial}{\partial n}\left(\frac{p}{\varrho} + gz + \frac{1}{2}q^2\right)$$

$$= \frac{\partial}{\partial n}\left(\frac{P}{\varrho}\right). \quad (14.7)$$

The variation of the constant of Bernoulli's equation, normal to the streamlines, is thus related to the angular velocity of the element.

There are several cases of flows in which the total pressure P is the same for different streamlines.

One of these cases is illustrated in Fig. 14.3 and is of the flow of a fluid past a body, the flow being uniform well up-

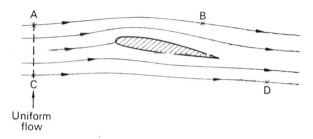

Fig. 14.3

stream. An observer travelling with the uniform flow would see it as a stationary fluid for which eqn. (9.20) gives

$$p + \varrho g z = \text{constant.}$$

The flow upstream being uniform, then at points such as those marked A and C in Fig. 14.3, the velocity will be the same. Thus for these two points,

$$p_A + \varrho g z_A + \tfrac{1}{2} q_A^2 = p_c + \varrho g z_c + \tfrac{1}{2} q_c^2$$
or
$$P_A = P_c,$$

so that, in the region of uniform flow, the total pressure is the same on all streamlines. The total pressure is also constant along streamlines on which the Bernoulli equation is valid. Points on such streamlines are marked B and D in Fig. 14.3. Thus

$$P_B = P_A = P_C = P_D,$$

and the total pressure is constant for all streamlines in the flow to which the Bernoulli equation is applicable. Thus throughout this region of the flow,

$$\frac{\partial P}{\partial n} = 0,$$

and so, from eqn. (14.7),

$$\omega = 0,$$

and this portion of the flow is irrotational.

A similar discussion shows that this result also applies to a flow that is developing out of a large reservoir in which the velocity far enough from the exit is effectively zero.

14.4 Viscous flow between plates

A type of flow, embodying some of the features already discussed, is illustrated in Fig. 14.4. This sketches the two-dimensional, incompressible, steady flow in the x–y plane between two flat parallel plates. Specifying that weight forces act in the z-direction, normal to the plane of the flow shown, simplifies the analysis whilst retaining its present usefulness.

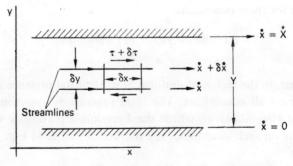

Fig. 14.4

THE MOMENTUM EQUATION FOR FLUIDS IN MOTION 239

Requiring that all the streamlines are straight and parallel to the x-direction gives the following results:

(a) Application of the continuity equation (12.5) means that $\dot{x}$ is independent of the value of x. Thus
$$\dot{x} = \dot{x}(y). \tag{14.8}$$

(b) The radius of curvature r being infinite, then from eqn. (11.10)
$$\tau = \mu \frac{\partial \dot{x}}{\partial y}, \tag{14.9}$$

and so, noting eqn. (14.8) and taking μ as a constant,
$$\tau = \tau(y). \tag{14.10}$$

(c) The shear stresses on the two vertical faces of the element sketched in Fig. 14.4 being equal and opposite, it follows from the discussion of § 13.3 that eqn. (14.6) is applicable, so that r being infinite,
$$\frac{\partial p}{\partial y} = 0$$
and so
$$p = p(x). \tag{14.11}$$

The force on the element in the x-direction is, by study of Fig. 14.4, given by
$$-\frac{dp}{dx} \delta x \, \delta y + (\tau + \delta \tau) \, \delta x - \tau \, \delta x.$$

Equation 14.8 indicates that there is no acceleration in the x-direction and so this force is equal to zero. Furthermore, noting eqn. (14.10) enables the increment in τ to be expressed by
$$\delta \tau = \frac{d\tau}{dy} \delta y$$

so that
$$\frac{dp}{dx} = \frac{d\tau}{dy}. \tag{14.12}$$

Equation 14.11 shows that the left-hand side of this relation is a function of x only and eqn. (14.10) shows that the right-hand side is a function of y only. These conditions can only be satisfied if both sides are a constant. Thus integration with respect to y gives

$$B + y\frac{dp}{dx} = \tau$$

and, substituting from eqn. (14.9), leads to

$$\mu\frac{d\dot{x}}{dy} = B + y\frac{dp}{dx}.$$

Integrating again, then,

$$\mu\dot{x} = A + By + \frac{1}{2}\frac{dp}{dx}y^2, \tag{14.13}$$

showing that the velocity varies parabolically.

If the lower plate is stationary so that when $y = 0, \dot{x} = 0$, then $A = 0$. If the upper plate, at a height Y, is moving at a speed $\dot{X}$, so that when $y = Y, \dot{x} = \dot{X}$, then

$$\mu\dot{X} = BY + \frac{1}{2}\frac{dp}{dx}Y^2.$$

Thus eqn. (14.13) becomes

$$\frac{\dot{x}}{\dot{X}} = \frac{y}{Y} - \frac{1}{2}\frac{Y^2}{\mu\dot{X}}\frac{dp}{dx}\left(\frac{y}{Y} - \frac{y^2}{Y^2}\right).$$

At the lower plate, where $y = 0$, the velocity gradient becomes

$$\mu\left(\frac{d\dot{x}}{dy}\right)_0 = \frac{\mu\dot{X}}{2Y}\left[2 - \frac{Y^2}{\mu\dot{X}}\frac{dp}{dx}\right]. \tag{14.14}$$

The non-dimensional quantity, $\dfrac{Y^2}{\mu \dot{X}} \dfrac{dp}{dx}$, has a close analogy for the flow in a boundary layer where it is called Polhausen's parameter. The velocity profile is dependent upon its value; three profiles are sketched in Fig. (14.5).

When the parameter is zero, a straight-line profile is obtained; when it is equal to 2, the slope and hence the shear stress

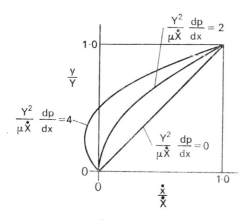

Fig. 14.5

at the lower plate is zero; and when it is greater than 2 there is a reversal of the flow direction above the lower plate. This is analogous to the phenomenon of boundary layer separation which was described in § 11.13 in which the dragging forward of the flow by viscous stresses is opposed by a positive pressure gradient until the latter reaches the level where a flow reversal takes place and the boundary layer leaves the surface.

When the pressure gradient is zero the shear stress on the lower plate is, from eqn. (14.14),

$$\tau_0 = \frac{\mu \dot{X}}{Y}.$$

A non-dimensional friction coefficient C_f can be defined by

$$C_f \equiv \frac{\tau_0}{\frac{1}{2}\varrho \dot{X}^2}$$

and substitution for τ_0 gives

$$C_f = \frac{2\mu}{\varrho \dot{X} Y}.$$

The non-dimensional quantity $\varrho \dot{X} Y / \mu$ is called the Reynolds number. This and the Polhausen parameter, two important quantities in the motion of fluids, are discussed in later volumes in the same series.[†]

For the purposes of § 15.7, the work done upon an element of fluid by the shear stresses is now evaluated. Referring again to Fig. 14.4, the rate at which work is being done upon the upper surface of the element is the product of the force, $(\tau + \delta\tau)\,\delta x$, and the velocity, $\dot{x} + \delta\dot{x}$, that is,

$$(\tau + \delta\tau)(\dot{x} + \delta\dot{x})\,\delta x.$$

Similarly, upon the lower surface the work is being done at a rate

$$-\tau \dot{x}\,\delta x,$$

where the negative sign arises because of the reversed direction of the shear stress. The net rate is thus

$$\begin{aligned}(\tau + \delta\tau)(\dot{x} + \delta\dot{x})\,\delta x - \tau\dot{x}\,\delta x &= (\tau\,\delta\dot{x} + \dot{x}\,\delta\tau)\,\delta x \\ &= \delta(\tau\dot{x})\,\delta x \\ &= \frac{d}{dy}(\tau\dot{x})\,\delta y\,\delta x.\end{aligned}$$

[†] Bradshaw, *op. cit.*

THE MOMENTUM EQUATION FOR FLUIDS IN MOTION 243

Substituting from eqns. (14.9) and (12.5), the rate of doing work divided by the rate of mass flow is

$$\frac{\dot{W}\tau}{\delta \dot{m}} = \frac{(d/dy)[\mu\dot{x}(d\dot{x}/dy)]\,\delta x\,\delta y}{\varrho \dot{x}\,\delta y}$$

$$= \frac{\mu}{\varrho \dot{x}} \frac{d^2}{dy^2}\left(\frac{1}{2}\dot{x}^2\right)\delta x. \qquad (14.15)$$

14.5 Flow through a control volume

Figure 14.6 is a sketch of a control volume, drawn in a fluid in motion, and so fixed in an x, y, z coordinate system that the weight force is in the negative z-direction. An element of

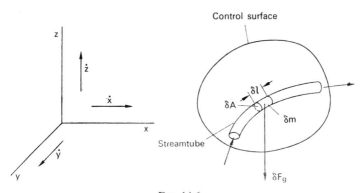

Fig. 14.6

fluid, of mass δm, forming a length δl of a streamtube that extends across the control volume, is shown sketched in this figure.

The z-component of the force upon the element δF_z is composed of three parts; that due to pressure forces δF_{zp}, that due to shear forces $\delta F_{z\tau}$, and that due to weight forces δF_g.

Applying Newton's law, eqn. (1.8), to the element and in the

z-direction, gives

$$\delta F_z = \delta m \frac{D\dot{z}}{Dt}.$$

Limiting discussion to steady flow and using eqn. (11.6) this becomes

$$\delta F_z = \delta m \, q \, \frac{\partial \dot{z}}{\partial l}.$$

Successive substitution of eqns. (12.2) and (12.5) changes this relation to

$$\delta F_z = \varrho \, \delta A \, \delta l \, q \, \frac{\partial \dot{z}}{\partial l}$$
$$= \delta \dot{m} \, \delta \dot{z}. \qquad (14.16)$$

This equation can be written down for every element within the control volume and then summed for all these elements. The left-hand side can be expanded to its three components becoming

$$\delta F_z = \delta F_{zp} + \delta F_{z\tau} + \delta F_g.$$

When the first of these three components is summed, all the pressure forces on every element cancel with the equal and opposite reactions on every other adjacent one except that the forces acting around the surface of the control volume are left to give a force F_{zp} as the result of this summation.[†] Expressed mathematically, this is

$$\Sigma \, \delta F_{zp} = F_{zp}.$$

The same argument applies to the summation of the shear forces, so that

$$\Sigma \, \delta F_{z\tau} = F_{z\tau}.$$

In contrast the gravitational forces are all additive; they do not cancel in any way. The force on each element is

$$\delta F_g = -g \, \delta m$$

† See the similar discussion of § 6.4.

THE MOMENTUM EQUATION FOR FLUIDS IN MOTION 245

and, for g being a constant,

$$\Sigma \, \delta F_g = -g \Sigma \, \delta m$$
$$= -gm,$$

where m is the mass of fluid contained within the control volume.

Thus the left-hand side of eqn. (14.16) sums to

$$F_{zp} + F_{z\tau} - gm.$$

To sum the right-hand side of eqn. (14.16) a double integration over the whole control volume is performed as

$$\iint d\dot{m} \, d\dot{z}.$$

This is done by first summing along the streamtube and then summing for all streamtubes. But along a streamtube, by the continuity equation, $d\dot{m}$ is a constant for steady flow and so

$$\iint d\dot{m} \, d\dot{z} = \int d\dot{m} \int d\dot{z}$$
$$= \int d\dot{m}(\dot{z}_{\text{out}} - \dot{z}_{\text{in}})$$
$$= \int_{\text{out}} \dot{z} \, d\dot{m} - \int_{\text{in}} \dot{z} \, d\dot{m},$$

where the suffixes *out* and *in* respectively indicate conditions at outlet from and inlet to the control volume.[†]

Thus the momentum equation, in the z-direction, for a control volume is

$$F_{zp} + F_{z\tau} - gm = \int_{\text{out}} \dot{z} \, d\dot{m} - \int_{\text{in}} \dot{z} \, d\dot{m}. \qquad (14.17)$$

Similar expressions for the forces in the x- and y-directions are derived in an identical manner except that the weight force has no component in these directions. They are

$$F_{xp} + F_{x\tau} = \int_{\text{out}} \dot{x} \, d\dot{m} - \int_{\text{in}} \dot{x} \, d\dot{m}, \qquad (14.18)$$

$$F_{yp} + F_{y\tau} = \int_{\text{out}} \dot{y} \, d\dot{m} - \int_{\text{in}} \dot{y} \, d\dot{m}. \qquad (14.19)$$

[†] This conversion from a volume integral to a surface integral is an elementary example of Green's theorem.

It is left as an exercise to show that, by using eqn. (12.3), the analysis can be repeated for unsteady flow so that eqn. (14.17) is just changed by adding the extra term $\iint \frac{\partial}{\partial t} (\dot{z}\, dm)$ to the right-hand side.

The three equations, (14.17), (14.18), and (14.19), have been obtained for the control volume by applying eqn. (1.8) which is only valid for a rigid system. If eqn. (1.10) is applied directly to the flow in a control volume, then equations identical to these and including the unsteady flow terms can be derived.[1] This shows that eqn. (1.10) is valid for a control volume of varying mass.

14.6 Physical significance of the wake

Application of the momentum equation for a control volume to the flow in a wake reveals the physical significance of the latter. This is illustrated in Fig. 14.7 which is a sketch of the two-dimensional flow past an aerofoil shape around which a control volume, marked $ABCD$, is drawn in an x–y coordinate system.

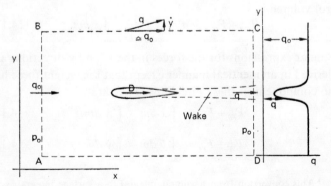

Fig. 14.7

THE MOMENTUM EQUATION FOR FLUIDS IN MOTION

Sufficiently far upstream the pressure and velocity are uniform along AB at the values denoted by p_0 and q_0; sufficiently downstream the pressure is again uniform at a value p_0 but the velocity is reduced within the wake to a value q. A typical variation of q with y is given in this figure.

Along BC the flow will be deflected outwards and then inwards by the presence of the aerofoil. If BC is sufficiently far out then the deflection will be small and the x-component of the velocity can be taken as being equal to q_0.

Considering values per unit length perpendicular to the flow plane, then with a drag force D acting upon the aerofoil in the x-direction, the total force F_x, acting upon the control volume, is given by

$$F_x = p_0(y_B - y_A) - p_0(y_C - y_D) - D = -D.$$

The flow being steady, application of both the momentum and continuity equations, (14.18) and (12.5), then gives

$$F_x = \int_{DC} \varrho q^2 \, dy + \int_{BC} \varrho q_0 \dot{y} \, dx + \int_{AD} \varrho q_0 \dot{y} \, dx - \int_{AB} \varrho q_0^2 \, dy$$

which, from symmetry, becomes

$$F_x = \int_{DC} \varrho q^2 \, dy + 2 \int_{BC} \varrho q_0 \dot{y} \, dx - \int_{AB} \varrho q_c^2 \, dy.$$

Application of the continuity equation, (12.11), gives

$$\int_{AB} \varrho q_0 \, dy = \int_{DC} \varrho q \, dy + 2 \int_{BC} \varrho \dot{y} \, dx.$$

Multiplication of this equation by q_0 and substitution into the previous ones to eliminate $\dot{y}$ and F_x results in

$$D = \int_{DC} \varrho q(q_0 - q) \, dy. \tag{14.20}$$

This shows that the drag force is related to the deficit in velocity in the wake, that is $(q_0 - q)$.

This drag force is formed by integration, around the aerofoil surface, of both shear forces and pressure forces. That both types of stress are significant can be illustrated by the case of a thin, flat plate immersed in a flow. If the plate is normal to the oncoming stream, shear forces have no component in the drag direction and so all the drag is formed by pressure forces. With the plate aligned tangentially to the flow the converse applies, all the drag is composed of frictional shear forces.

If, as described in § 14.3, frictional effects were absent from this flow, the Bernoulli equation would have a constant value throughout it. As a consequence, along CD, where the pressure has returned to its upstream value of p_0, then so also would the velocity return to a value of q_0; there would be no wake, and hence from eqn. (14.20) no drag. This study brings out the point that all the resistance in this steady two-dimensional flow, derived both from shear and from pressure forces acting on the body, exists because of the presence of viscous effects.

14.7 Unsteadiness in the flow past an aerofoil

In the previous application of the momentum equation to the flow past an aerofoil, it was assumed that the flow was steady. In fact all real flows are unsteady to some degree resulting in the drag force fluctuating with time. But if the control volume is drawn sufficiently far from the aerofoil so that the flow and fluid properties around it are effectively steady, and if the aerofoil itself is stationary so that the boundary conditions of the flow at it are steady,[†] then experimental evidence supports the use of the steady flow momentum equation for a control volume. The drag force obtained then corresponds to a time average value.

If, to the flow system illustrated in Fig. 14.7, is added a velocity q_0 in the negative x-direction, the flow is that due to the

[†] The velocity at a solid surface is zero, as explained in § 5.3.

THE MOMENTUM EQUATION FOR FLUIDS IN MOTION 249

motion of an aerofoil at a steady velocity q_0 through a stationary atmosphere. This flow is illustrated in Fig. 14.8. At a point, such as that marked A, a particle of fluid will move forwards, outwards, and inwards returning to point B after the aerofoil has passed.[2] The flow is now an unsteady one because conditions at fixed points such as C are varying with time.

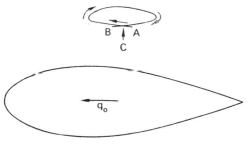

Fig. 14.8

Around the control volume (Fig. 14.7) the velocity is now $(q-q_0)$ along CD and has a zero x-component along $CBAD$. It is not now zero along that part of the control volume boundary formed by the aerofoil surface.† The contributions to the right-hand side of eqn. (14.18) are as follows. Along CD,

$$\int_{\text{out}} \dot{x} \, d\dot{m} = \int_{DC} \varrho(q-q_0)^2 \, dy.$$

Along $CBAD$,

$$\int_{\text{out}} \dot{x} \, d\dot{m} - \int_{\text{in}} \dot{x} \, d\dot{m} = 0.$$

The contribution from the aerofoil surface is illustrated in Fig. 14.9 where two small elements of the surface—one forward facing, one rearward facing—are shown. The x-velocity component of the forward-facing element is $-q_0$; the corre-

---

† This is analogous to the example of a piston described in § 12.5.

sponding mass flow rate into the control volume is $\varrho q_0 \, dy$ and so the contribution by this element is

$$(\dot{x} \, d\dot{m})_{\text{in}} = -\varrho q_0^2 \, dy.$$

Correspondingly the contribution by the rearward-facing element is

$$(\dot{x} \, d\dot{m})_{\text{out}} = -\varrho q_0^2 \, dy$$

and so these two terms cancel and there is no contribution from the aerofoil surface as a whole.

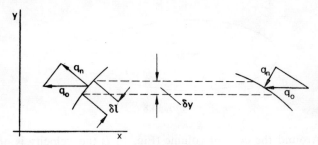

Fig. 14.9

Thus application of the momentum equation results in

$$-D = \int_{DC} \varrho(q-q_0)^2 \, dy + (UT), \qquad (14.21)$$

where (UT) denotes the extra term that arises in the momentum equation when the flow is unsteady. The drag results from the relative motion between the aerofoil and the fluid and thus is independent of the velocity of an observer. Thus, comparing this latter equation with eqn. (14.20) gives

$$(UT) = -\int_{DC} \varrho q(q_0 - q) \, dy - \int_{DC} \varrho(q - q_0)^2 \, dy$$
$$= \int_{DC} \varrho q_0 (q - q_0) \, dy.$$

When the flow is free of viscous effects the unsteady terms are zero. It is important to note that this unsteady flow is ob-

THE MOMENTUM EQUATION FOR FLUIDS IN MOTION 251

tained from a non-accelerating flow by addition of a velocity, and so the aerofoil is travelling at a steady speed. When it is accelerating, further terms exist in eqn. (14.21) so that the drag is changed, sometimes quite appreciably.[3] A drag can then exist even in the absence of viscous effects.

14.8 Angular momentum for a control volume

Referring again to Fig. (14.6), the element of mass δm has a radial ordinate r_x in the plane $x = $ constant. If the angular velocity of this radius is $\dot{\theta}_x$ then the acceleration of the element of fluid perpendicular to r_x is given by eqn. (1.6), so that the force upon the element in this direction is

$$\frac{\delta m}{r_x} \frac{\delta(r_x^2 \dot{\theta}_x)}{\delta t}.$$

The element of torque δN_x becomes

$$\delta N_x = \frac{\delta m}{\delta t} \delta(r_x^2 \dot{\theta}_x)$$
$$= \delta \dot{m}\, \delta(r_x^2 \dot{\theta}_x),$$

where, as before, $\delta \dot{m}$ is the rate of mass flow along the stream tube. To obtain the total torque about the x-axis N_x that is acting upon the fluid in the control volume this expression is summed in the manner used in § 14.5. That is, for steady flow,

$$\begin{aligned} N_x &= \iint d\dot{m}\, d(r_x^2 \dot{\theta}_x) \\ &= \int d\dot{m} \int d(r_x^2 \dot{\theta}_x) \\ &= \int d\dot{m}[(r_x^2 \dot{\theta}_x)_{\text{out}} - (r_x^2 \dot{\theta}_x)_{\text{in}}] \\ &= \int_{\text{out}} r_x^2 \dot{\theta}_x\, d\dot{m} - \int_{\text{in}} r_x^2 \dot{\theta}_x\, d\dot{m}. \end{aligned} \quad (14.22)$$

It is left as an exercise to show, by arguments analogous to those of § 14.5, that this torque is composed of the torque due to pressure and shear forces acting around the control surface plus the torque due to the weight of the fluid within the control volume acting at the y-ordinate of its centre of gravity.

14.9 Pumping by a ducted fan

Figure 14.10 is a sketch of a fan that is pumping fluid along a duct whose axis is taken as the x-axis. The discussion can be simplified by specifying that the flow is incompressible, that viscous effects are significant only within boundary layers and wakes that are thin, and that a control surface is drawn so as to form a streamtube of constant cross-sectional area that just excludes the boundary layer upon the duct wall.

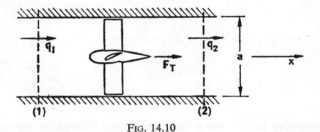

Fig. 14.10

The major portion of the flow between stations such as those indicated (1) and (2) in Fig. 14.10 will be free of viscous effects. If the Bernoulli equation, (14.4), is applied to this flow, then $P_1 = P_2$, and the total pressure is unchanged through the fan. This is wrong; the whole object of a fan is to increase the total pressure by increasing any one or more of the pressure, the velocity, and the height of the fluid.

It is obvious that an error arises in applying the Bernoulli equation. The assumptions made are valid; the point overlooked is that this flow is unsteady. Only by being so can the

THE MOMENTUM EQUATION FOR FLUIDS IN MOTION 253

total pressure be changed in a flow that is free of net viscous forces.

As shown in Fig. 14.10, the fluid will be flowing out of the control volume across section (2) where, due to the torque exerted by the fan upon the fluid, the flow will have a swirling component of the velocity so that the axial component will be $\dot{x}_2$. Choosing the two planes, (1) and (2) of Fig. 14.10, first to be sufficiently far up and downstream so that the flow conditions there are steady, and second to be perpendicular to the x-axis so that the shear forces acting on them have no x-component, and denoting by F_T the force on the fan, then the force on the control volume F_x is

$$F_x = \int_1 p\, da - \int_2 p\, da + F_T$$
$$= \int (p_1 - p_2)\, da + F_T.$$

Applying the momentum equation, (14.18), and noting eqn. (12.5), gives

$$F_x = \int_2 \dot{x}\, d\dot{m} - \int_1 \dot{x}\, d\dot{m} + (UT)$$
$$= \int \varrho(\dot{x}_2^2 - \dot{x}_1^2)\, da + (UT),$$

where (UT) denotes the unsteady flow terms in the momentum equation.[†] Thus

$$\int (p_1 - p_2)\, da + F_T = \int \varrho(\dot{x}_2^2 - \dot{x}_1^2)\, da + (UT). \quad (14.23)$$

The whole flow can be reduced to a steady one by applying to the system an angular velocity ω about the duct axis which brings the fan to rest. The boundary conditions then become steady everywhere; on the wall of the duct they are ones of constant velocity at a fixed position. Doing this introduces no

---

[†] During the derivation of the momentum equation for unsteady flow, the first term on the right-hand side is obtained from an instantaneous integration along a streamtube. Thus it does not matter that the streamtubes change with time because the flow is unsteady.

change in $\dot{x}$ anywhere and so there is no change in the term $(\dot{x}_2^2 - \dot{x}_1^2)$ in eqn. (14.23).

Furthermore, if there is no change in $\dot{x}$ there is no change in the x-component of the acceleration upon any element, and hence there is no change in the x-component of the force on an element δF_x.

Adding the angular velocity ω makes no change in the relative velocity between elements and so makes no change in the shear stress distribution and hence in the shear force upon an element.

Now

$$\delta F_x = -\frac{\partial p}{\partial x}\,\delta V + \text{shear force},$$

and so there is no change in $\partial p/\partial x$. Pressure changes in the x-direction being obtained by integration of $\partial p/\partial x$ are thus unchanged and so there is no alteration of the term $(p_1 - p_2)$ in eqn. (14.23).

There being no change in the shear stresses and in the differences of pressure, there is no change in F_T. So that finally, application of eqn. (14.18) to the steady flow gives

$$\int (p_1 - p_2)\,da + F_T = \int \varrho(\dot{x}_2^2 - \dot{x}_1^2)\,da.$$

Comparison with eqn. (14.23) shows that

$$(UT) = 0. \tag{14.24}$$

This result, that the unsteady flow terms in the control volume momentum equation are zero under the conditions specified seems to be true for fluid machines in general.[4][†]

This result is not, however, true for the Bernoulli equation which, for this incompressible unsteady flow, becomes

$$p_1 + \tfrac{1}{2}\varrho q_1^2 = p_2 + \tfrac{1}{2}\varrho q_2^2 + (UT) \tag{14.25}$$

---

† A closely similar demonstration that the unsteady terms in the control volume angular momentum equation are zero is left as an exercise.

THE MOMENTUM EQUATION FOR FLUIDS IN MOTION   255

For a simple example, in which the pressure and velocity is assumed uniform at each station and in which $q_1 = \dot{x}_1$ and $q_2 \simeq \dot{x}_2$, the continuity equation, (12.11), gives that

$$\dot{x}_1 = \dot{x}_2.$$

Equation 14.23 together with eqn. (14.24) becomes

$$p_1 a - p_2 a + F_T = 0$$

so that
$$p_2 - p_1 = F_T/a, \qquad (14.26)$$

showing that the pressure rise across the fan is equal to the thrust per unit area. Equation (14.25) then gives

$$(UT) = F_T/a.$$

which is not zero.

14.10 Momentum equation for the flow with diffusion

Each particle within a small volume δV will have a drift velocity of v_i. The corresponding acceleration will be Dv_i/Dt. The product of the mass and the acceleration will be $M_i A_m \times (Dv_i/Dt)$.

The total force δF upon the volume δV will equal the summation of all these particle accelerations. This summation is, following the analysis of § 12.8,

$$\delta F = \sum M_i A_m \frac{Dv_i}{Dt}$$

$$= \sum \frac{D}{Dt}(M_i A_m v_i)$$

$$= \frac{D}{Dt} \sum (M_i A_m v_i)$$

$$= \frac{D}{Dt} \sum_i M_i A_m v_i N_i$$

$$= \frac{D}{Dt} \sum_i \varrho v_i \delta V.$$

Substitution from eqn. (12.29) gives

$$\delta F = \frac{D}{Dt}(\varrho \dot{x}\, \delta V).$$

This analysis is made by considering a fixed identity of particles so that

$$\varrho\, \delta V = \text{constant}$$
$$= \delta m.$$

Thus

$$\delta F = \varrho\, \delta V \frac{D\dot{x}}{Dt}$$
$$= \delta m \frac{D\dot{x}}{Dt}.$$

This relation is identical in form to the acceleration term of § 14.1 and to the first equation of § 14.5. It follows that the momentum equation for a flow in which diffusion is present is identical in form to those of eqns. (14.3), (14.17), (14.18), and (14.19) provided that the velocities are defined by eqn. (12.29).

References

1. SHAPIRO, A. H., *The Dynamics and Thermodynamics of Compressible Fluid Flow*, Ronald, New York, 1953, vol. 1, p. 16.
2. DURAND, W. F., *Fluid Mechanics*, pt. 1, VII, 1, 2, p. 186, *Aerodynamic Theory* (ed. W. F. Durand), vol. 1, Springer, Berlin, 1934.
3. MUNK, M. M., *Fluid Mechanics*, pt. 1, C, III, 2, p. 242. *Aerodynamic Theory* (ed. W. F. Durand), vol 1, Springer, Berlin, 1934.
4. PRESTON, J. H., The non-steady irrotational flow of an inviscid, incompressible fluid, with special reference to changes in total pressure through flow machines, *Aero. Quart.* **12**, 343 (May 1961).

CHAPTER 15

APPLICATION OF THE FIRST LAW OF THERMODYNAMICS TO FLUIDS IN MOTION

15.1 Characteristics of the infinitesimal element

In §§ 7.2 and 8.13 an infinitesimal element of a solid was seen to satisfy the requirements of the experiments of Joule upon which was based the formulation of the first law of thermodynamics. The infinitesimal element of a fluid also satisfies these requirements. Additionally, as described in § 8.13, thermal and mechanical processes upon an infinitesimal element can be so summed that they proceed simultaneously.

15.2 Work done by pressures on a moving element of fluid

Figure 15.1 is a sketch of an element of fluid composed of a length δl of a streamtube cut off by two planes normal to the flow direction. This element, of mass δm, is moving along a path-line so that in time δt it moves to the adjacent position shown dotted in outline. At the upstream end of the element a pressure p is acting upon a surface area δA which is moving at a velocity q. Thus a force $p\delta A$ is doing work to the element and relative to the space axes at a rate $p\delta A q$. Making use of eqn. (12.5) this rate of doing work is expressed by

$$\frac{p}{\varrho} \delta \dot{m}. \qquad (15.1)$$

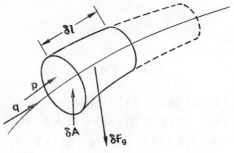

Fig. 15.1

In a similar way the pressure acting on the downstream face is also doing work, but now its point of application is moving in a direction exactly opposite to the direction of the force and so the rate of doing work is negative. Except for this change of sign, the rate of doing work at the downstream face is equal to the rate at the upstream face, given by eqn. (15.1), plus an increment; this is expressed by

$$-\left[\frac{p}{\varrho}\,\delta\dot{m} + \delta\left(\frac{p}{\varrho}\,\delta\dot{m}\right)\right]. \tag{15.2}$$

The pressures acting upon the side walls of the streamtube do no work because their point of application is moving perpendicularly to their direction.[†]

Thus the total rate at which the pressures are doing work to the element is given by the sum of eqns. (15.1) and (15.2), that is

$$\frac{p}{\varrho}\,\delta\dot{m} - \left[\frac{p}{\varrho}\,\delta\dot{m} + \delta\left(\frac{p}{\varrho}\,\delta\dot{m}\right)\right]$$
$$= -\delta\left(\frac{p}{\varrho}\,\delta\dot{m}\right).$$

† This follows from the definitions of a pressure, of a streamtube and of work.

LAW OF THERMODYNAMICS TO FLUIDS IN MOTION   259

Specifying steady flow first makes the streamline and path-line coincident, and second makes $\delta \dot{m}$ a constant so that it can be taken outside the differential. Thus the rate of doing work to the element is

$$-\delta \dot{m}\, \delta\left(\frac{p}{\varrho}\right) = -\frac{\delta m}{\delta t}\, \delta\left(\frac{p}{\varrho}\right).$$

And so, the rate of doing work per mass unit is

$$-\frac{1}{\delta t}\, \delta\left(\frac{p}{\varrho}\right)$$

and the work done per mass unit, in time δt is,

$$-\delta\left(\frac{p}{\varrho}\right). \tag{15.3}$$

This element is being both changed in volume and translated, so the process just considered is not that discussed in § 8.3, where only the former type is analysed.[†] This can be illustrated by considering separately the effect of these two types of process, the discussion being a simple extension of that of § 8.13. The system is here an infinitesimal element for which p is uniform throughout. If its volume change during the infinitesimal process is δV, then eqn. (8.5) for the work done as a result of this volume change is

$$W_{\text{to}} = -p\, \delta V$$

The work done per mass unit is then

$$-\frac{p\, \delta V}{\delta m}$$

$$= -p\, \delta\left(\frac{V}{\delta m}\right)$$

$$= -p\, \delta\left(\frac{1}{\varrho}\right) \tag{15.4}$$

[†] No work is done in the mechanical process of translating an object if there is no frictional resistance, no body force acting, and no change in kinetic energy.

where ϱ is the density. This is the work done in the absence of translation.

The component of the force on the element, of volume V, in the direction of motion is

$$-\frac{\partial p}{\partial l} V.$$

The work done in moving a distance δl is thus

$$-\frac{\partial p}{\partial l} V \, \delta l$$

and the work of translation per mass unit is

$$-\frac{(\partial p/\partial l)V \, \delta l}{\varrho V} = -\frac{1}{\varrho} \, \delta p. \tag{15.5}$$

The total work done is the sum of eqns. (15.4) and (15.5) which is

$$-p \, \delta\left(\frac{1}{\varrho}\right) - \frac{1}{\varrho} \, \delta p = -\delta\left(\frac{p}{\varrho}\right)$$

which, as it should be, is the result given by eqn. (15.3).

In the special case of the motion of an incompressible fluid the term (15.4) is then zero but the term (15.5) due to translation of the element remains.

15.3 Work done on a moving fluid element by gravity forces

As shown in Fig. 15.1 a weight force δF_g of amount $g \delta m$ acts upon the element in a vertically downward direction. If, in a time δt, the element moves vertically upwards a distance δz relative to the chosen system of axes, then the work done by this force is $-g \, \delta m \, \delta z$ and so the work done to the element per mass unit is

$$-g \, \delta z \tag{15.6}$$

15.4 Work done by shear stresses

As illustrated in § 11.11, work can be done upon an infinitesimal element by the tangential shear stresses. The rate of doing work in this way, to the element per mass unit, is written as

$$\frac{\dot{W}_{\tau\,\text{to}}}{\delta m}$$

and the work done in time δt is then,

$$\frac{\dot{W}_{\tau\,\text{to}}}{\delta m}\,\delta t = \frac{\dot{W}_{\tau\,\text{to}}}{\delta \dot{m}}. \tag{15.7}$$

15.5 Heat applied to the element

With temperature gradients present within the fluid, heat can be applied to the element by the surrounding ones as described for the case of the solid in §§ 7.2 and 8.13. Calling this rate of heating $\dot{Q}_{\text{to}}$ then, as in the discussion of the previous section, the heat per mass unit applied in time δt is

$$\frac{\dot{Q}_{\text{to}}}{\delta \dot{m}}. \tag{15.8}$$

15.6 The energy change of an element

The element being infinitesimal in size, it has a kinetic energy of $\frac{1}{2}q^2$ per mass unit and an internal energy of e per mass unit, both of which are single valued. Then, by extension of eqn.

262   THERMOMECHANICS

(8.29), a quantity e_T can be defined by

$$e_T \equiv \frac{E_T}{\delta m}$$

$$= e + \tfrac{1}{2} q^2. \tag{15.9}$$

The rate at which this is changing is De_T/Dt and so the change in a time δt is

$$\frac{De_T}{Dt} \delta t. \tag{15.10}$$

15.7 The energy equation for steady flow along a streamline

As was done in § 8.13, two infinitesimal processes, one according to the first law of thermodynamics, the other to Newton's law, are now summed. Using the items (15.3), (15.6), (15.7), (15.8), and (15.10)[†] this leads to

$$-\delta\left(\frac{p}{\varrho}\right) - g\,\delta z + \frac{\dot{W}_{\tau\text{to}}}{\delta \dot{m}} + \frac{\dot{Q}_{\text{to}}}{\delta \dot{m}} = \frac{De_T}{Dt}\delta t. \tag{15.11}$$

This relation can be integrated to express the changes that take place as the element moves over a finite distance. With g constant, the first two terms integrate to give, between positions 1 and 2 say,

$$\frac{p_1}{\varrho_1} - \frac{p_2}{\varrho_2} + gz_1 - gz_2.$$

If the flow is steady, then the discussion of § 11.3 shows that summation of the third term along the streamtube surface can be an instantaneous one, in the manner described in § 5.11.

---

† Other terms can arise: for instance magnetic, electric field, and surface tension forces can do work when a fluid is in motion and electricity can be applied to a conducting fluid. At the present stage such terms are excluded.

The streamtube end surfaces being perpendicular to the flow direction, the velocity will have no component in the tangential stress direction and so there will be no contribution to $\dot{W}_{\tau\text{to}}$ there. Thus with $\delta\dot{m}$ being a constant along the streamtube for this steady flow, the third term sums to give

$$\sum_{1}^{2} \frac{\dot{W}_{\tau\text{to}}}{\delta\dot{m}} = \frac{1}{\delta\dot{m}} \sum_{1}^{2} \dot{W}_{\tau\text{to}}.$$

A similar discussion applies to the summation of the fourth term, except that in this case there is a contribution at the ends of the streamtube when there are temperature gradients there that are relative to the fluid and in the flow direction.† Thus the fourth term sums to become

$$\sum_{1}^{2} \frac{\dot{Q}_{\text{to}}}{\delta\dot{m}} = \frac{1}{\delta\dot{m}} \sum_{1}^{2} \dot{Q}_{\text{to}}.$$

The right-hand side of eqn. (15.11) integrates as

$$\int_{1}^{2} \frac{De_T}{Dt}\, dt = \int_{1}^{2} De_T$$

$$= e_{T2} - e_{T1}.$$

There are two points of interest concerning these summations. First, all summations are with respect to time, and so the difficulties discussed in § 1.7 disappear. Second, the pressure term can be integrated without the necessity of specifying the nature of the process: this provides a marked contrast with the case of the Bernoulli momentum equation and so makes the energy relation the more generally useful of the two.

† The term $\dot{Q}_{\text{to}}$ refers to a happening relative to the element. At the entrance to the streamtube, the element that is under discussion is being followed by another element that may have a different temperature. At this instant in time there is then a contribution to $\dot{Q}_{\text{to}}$ between these two elements.

Collecting these terms, the energy equation for steady flow along a streamline becomes

$$\frac{p_1}{\varrho_1} - \frac{p_2}{\varrho_2} + gz_1 - gz_2 + \frac{1}{\delta \dot{m}} \sum_1^2 \dot{W}_{\tau \mathrm{to}}$$

$$+ \frac{1}{\delta \dot{m}} \sum_1^2 \dot{Q}_{\mathrm{to}} = e_{T2} - e_{T1}. \quad (15.12)$$

As discussed in § 11.3, this relation also provides an instantaneous comparison between any two points on a streamline.

The significance of the terms in $\dot{W}_{\tau \mathrm{to}}$ and $\dot{Q}_{\mathrm{to}}$ is now considered.

If was shown in § 11.11 that only in the case of solid body rotation is the $\dot{W}_{\tau \mathrm{to}}$ term identically zero. The general significance of this term can be illustrated by the viscous flow between parallel plates that was discussed in § 14.4. Using eqn. (14.15) and forming the summation of eqn. (15.12) gives

$$\frac{1}{\delta \dot{m}} \sum \dot{W}_{\tau \mathrm{to}} = \frac{\mu}{\varrho \dot{x}} \frac{\partial^2}{\partial y^2} \left(\frac{1}{2} \dot{x}^2\right) \int dx$$

$$= \frac{\mu x}{\varrho \dot{x}} \cdot \frac{\partial^2}{\partial y^2} \left(\frac{1}{2} \dot{x}^2\right) + \text{constant}. \quad (15.13)$$

An expression for the term in $\dot{Q}_{\mathrm{to}}$ can be obtained by a similar discussion. Noting eqn. (7.2) and applying an analysis very similar to that in § 14.4, the contribution to $\dot{Q}_{\mathrm{to}}$ that is applied to the element sketched in Fig. 14.4 is

$$-\frac{\partial \dot{Q}}{\partial x} \delta x$$

in the x-direction and is

$$-\frac{\partial \dot{Q}}{\partial y} \delta y$$

in the y-direction. Adding these terms, substituting from eqn. (7.1) with a similar relation for the y-direction, and taking k, the thermal conductivity, as constant, gives

$$\dot{Q}_{to} = k \left(\frac{\partial^2 T}{\partial x^2} + \frac{\partial^2 T}{\partial y^2} \right) \delta x \, \delta y.$$

Thus, using eqn. (12.5),

$$\frac{\dot{Q}_{to}}{\delta \dot{m}} = \frac{k}{\varrho \dot{x}} \left(\frac{\partial^2 T}{\partial x^2} + \frac{\partial^2 T}{\partial y^2} \right) \delta x.$$

The summation of this term in eqn. (15.12) thus becomes

$$\frac{1}{\delta \dot{m}} \sum \dot{Q}_{to} = \frac{k}{\varrho \dot{x}} \int \left(\frac{\partial^2 T}{\partial x^2} + \frac{\partial^2 T}{\partial y^2} \right) dx$$
$$= \frac{k}{\varrho \dot{x}} \left[\frac{\partial T}{\partial x} + \int \frac{\partial^2 T}{\partial y^2} \, dx \right] + \text{constant}.$$

For the particular case where $T = T(y)$ only[†] this becomes

$$\frac{1}{\delta \dot{m}} \sum \dot{Q}_{to} = \frac{kx}{\varrho \dot{x}} \frac{d^2 T}{dy^2} + \text{constant}. \qquad (15.14)$$

In studying the Bernoulli momentum equation in § 14.2, the numerical values of the various terms were discussed. Similarly, the terms of eqns. (15.13) and (15.14) are now compared with the others of eqn. (15.12).

Comparison between the $\dot{W}_{\tau to}$ term and the kinetic energy term, $\frac{1}{2} \dot{x}^2$, is made by rewriting eqn. (15.13) to give

$$\frac{1}{\frac{1}{2} \dot{x}^2} \frac{1}{\delta \dot{m}} \sum \dot{W}_{\tau to} = \frac{\mu}{\varrho \dot{X} Y} \left(\frac{\dot{X}}{\dot{x}} \right)^3 \frac{x}{Y} \frac{\partial^2}{\partial (y/Y)^2} \left(\frac{\dot{x}^2}{\dot{X}^2} \right) + \text{constant}. \qquad (15.15)$$

Similarly, a comparison between the $\dot{Q}_{to}$ term and the internal energy term e is made by using eqns. (8.17) and (9.29) or (9.31) to rewrite eqn. (15.14) to give

$$\frac{1}{e} \frac{1}{\delta \dot{m}} \sum \dot{Q}_{to} = \frac{k}{C_u \mu} \frac{\mu}{\varrho \dot{X} Y} \frac{\dot{X}}{\dot{x}} \frac{T_0}{T} \frac{x}{Y} \frac{\partial^2}{\partial (y/Y)^2} \left(\frac{T}{T_0} \right) + \text{constant}, \qquad (15.16)$$

where T_0 is a reference temperature such as that at $y = 0$.

[†] Details of this particular case are left to further study.[1]

The non-dimensional quantity $(C_u\mu)/k$ is a property and has a value of the order of 1·0 for air and 10 for water.† The non-dimensional quantity $\rho \dot{X} Y/\mu$ is an example of the Reynolds number mentioned in § 14.4.

Inspection of eqns. (15.15) and (15.16) shows that as either the velocity or the size of a flow pattern is changed, the Reynolds number changes but the non-dimensional quantities of eqns. (15.15) and (15.16), which are $\dot{x}/\dot{X}$, x/Y, y/Y, and T/T_0, remain unaffected. Thus the relative values of the terms $\dot{W}_{\tau \text{to}}$ and $\dot{Q}_{\text{to}}$ are controlled by the value of the Reynolds number. A value of 10^5 is small for a typical Reynolds number and then, for these terms to be significant in eqn. (15.12), respectively, the second differentials of the velocity and the temperature must be very large. This is usually only so within thin boundary layers and wakes with large heat rates; outside these regions these terms can usually be neglected.‡

It is important to note that as the speed of a process is reduced, so that correspondingly the Reynolds number gets smaller, then the friction and heat terms increase in significance. For example, if a piston is compressing the gas in a cylinder, then as the speed of the piston is reduced so more and more work is done by frictional stresses.§

Thus for many flows of interest eqn. (15.12) becomes

$$\frac{p_1}{\varrho_1} - \frac{p_2}{\varrho_2} + gz_1 - gz_2 = e_{T2} - e_{T1} \qquad (15.17)$$

---

† This quantity is closely analogous to one known as the Prandtl number which is defined as $(C_h\mu)/k$, where C_h is defined later in § 15.8.

‡ An example of a large value of the temperature gradient in the streamwise direction occurs in the flow through a shock wave.[2]

§ In the terms of the studies of later volumes the process becomes less like a reversible one. Another approach to this result is by a consideration of the physical significance of the Reynolds number. (See Bradshaw, *op. cit.*)

which for a two-property fluid can be rewritten as

$$\left(\frac{p}{\varrho}+u\right)_1 - \left(\frac{p}{\varrho}+u\right)_2 = \left(gz+\frac{1}{2}q^2\right)_2 - \left(gz+\frac{1}{2}q^2\right)_1. \quad (15.18)$$

15.8 Enthalpy

From the discussion of § 5.9 it follows that the quantity that appears on the left-hand side of eqn. (15.18) is an additive property per mass unit. It is called the enthalpy h so that

$$h \equiv \frac{p}{\varrho}+u. \quad (15.19)$$

Equation 15.18 can now be rewritten in the form

$$h_1 - h_2 = (gz+\tfrac{1}{2}q^2)_2 - (gz+\tfrac{1}{2}q^2)_1. \quad (15.20)$$

The enthalpy coefficient C_h is defined by[†]

$$C_h \equiv \left(\frac{\partial h}{\partial T}\right)_p.$$

In the case of a gas satisfying the relations (9.11) and (9.31), eqn. (15.19) becomes

$$h = \left(\frac{R}{M}+C_u\right)T$$

and then, by differentiating this expression,

$$C_h = \frac{R}{M}+C_u, \quad (15.21)$$

and thus as C_h is a constant, in this case

$$h = C_h T. \quad (15.22)$$

[†] This is usually called the coefficient of specific heat at constant pressure and denoted C_p (see discussion in § 8.8) Then care must be taken to avoid confusion with the use of C_p for pressure coefficient. (See Bradshaw, *op. cit.*)

Finally, for a flow of this ideal gas, and for which the gz term is negligible, eqn. (15.20) becomes

$$C_h(T_1-T_2)+\tfrac{1}{2}(q_1^2-q_2^2) = 0. \qquad (15.23)$$

15.9 The energy equation for the flow through a control volume

Figure 15.2 is a sketch of a control volume through which fluid is flowing. It is composed of elements of fluid, one of which is shown shaded within its streamtube.

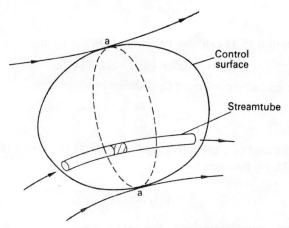

Fig. 15.2

Equation (15.12) can be multiplied throughout by $\delta \dot{m}$ and written down for each streamtube, then every version can be summed for all the streamtubes comprising the control volume. This is the procedure already used in § 12.2.

The first term, when so summed, can be expressed as

$$\sum \frac{p_1}{\varrho_1} \delta\dot{m} = \int_{\text{in}} \frac{p}{\varrho} \, d\dot{m},$$

the integration being performed over the control surface to the left of a–a, as indicated in Fig. 15.2. Similarly, the second term becomes

$$\int_{\text{out}} \frac{p}{\varrho} \, d\dot{m}.$$

In the same way, the third and fourth terms together become

$$\sum gz_1 \, \delta\dot{m} - \sum gz_2 \, \delta\dot{m} = \int_{\text{in}} gz \, d\dot{m} - \int_{\text{out}} gz \, d\dot{m}.$$

The quantity $\dot{W}_{\tau\,\text{to}}$ is the rate at which work is being done to the element by the forces imposed by the adjacent elements. There being no discontinuity in either stress or velocity at the boundaries between elements $\dot{W}_{\tau\,\text{to}}$ for any element will cancel with the corresponding terms to the adjacent ones. Thus if $\dot{W}_{\tau\,\text{to}}$ is summed along a streamline as in eqn. (15.12) and then summed for all streamlines, its total value will be the uncancelled part around the outer surfaces of the control volume $\dot{W}_{\tau s\,\text{to}}$ say. This summation can be expressed by

$$\sum \sum_1^2 \dot{W}_{\tau\,\text{to}} = \dot{W}_{\tau s\,\text{to}}.$$

Using the result discussed in § 5.13 that $\dot{Q}$ has an equal and coincident reaction, an almost identical argument is valid for the summation, over the control volume, of the $\dot{Q}$ term in eqn. (15.12), and so

$$\sum \sum_1^2 \dot{Q}_{\text{to}} = \dot{Q}_{s\,\text{to}}.$$

The terms on the right-hand side of eqn. (15.12) sum in a manner identical to that used for the first four terms. Thus

$$\sum e_{T2} \, \delta\dot{m} - \sum e_{T1} \, \delta\dot{m} = \int_{\text{out}} e_T \, d\dot{m} - \int_{\text{in}} e_T \, d\dot{m}.$$

Collecting these results of the summations gives the energy equation for a control volume as

$$\dot{W}_{\tau s \text{ to}} + \dot{Q}_{s \text{ to}} = \int_{\text{out}} \left(\frac{p}{\varrho} + gz + e_T\right) d\dot{m}$$

$$- \int_{\text{in}} \left(\frac{p}{\varrho} + gz + e_T\right) d\dot{m}. \quad (15.24)$$

15.10 The rate equation for a control volume

At any instant in time the fluid contained within a control volume forms a system. The rate at which work is being done by the pressures around the control surface $\dot{W}_{\sigma s \text{ to}}$ is given by eqn. (5.10). Substituting into this equation from eqn. (12.6) and remembering that in the former q_n is positive outward from the control volume, gives

$$\dot{W}_{\sigma s \text{ to}} = -\int_s p q_n \, da$$

$$= -\int_s \frac{p}{\varrho} \, d\dot{m}$$

$$= \int_{\text{in}} \frac{p}{\varrho} \, d\dot{m} - \int_{\text{out}} \frac{p}{\varrho} \, d\dot{m}$$

which are the sixth and third terms in eqn. (15.24).

The rate at which the gravity body forces are doing work throughout the control volume is given by eqn. (5.11). Noting

LAW OF THERMODYNAMICS TO FLUIDS IN MOTION 271

that $\beta = g$ and that $q_\beta = -dz/dt$, this equation becomes

$$\dot{W}_{\beta \text{ to}} = -\int_v \beta q_\beta \, dm$$

$$= -g \int_v \frac{dz}{dt} \, dm$$

$$= -g \int_v dz \, d\dot{m}.$$

As before, this volume integration is firstly performed along a streamtube for which $d\dot{m} = $ constant and then by summing for all streamtubes. Thus

$$\dot{W}_{\beta \text{ to}} = -g \int d\dot{m} \int dz$$
$$= -g \int (z_{\text{out}} - z_{\text{in}}) \, d\dot{m}$$
$$= g \int_{\text{in}} z \, d\dot{m} - g \int_{\text{out}} z \, d\dot{m},$$

which are the seventh and fourth terms in eqn. (15.24).

The total energy content of each element is $e_T \, dm$. Its rate of change in a steady flow is

$$\frac{D}{Dt}(e_T \, dm)$$

$$= \delta m \frac{De_T}{Dt}$$

$$= \delta m q \frac{de_T}{dl}$$

$$= \delta m \frac{\delta l}{\delta t} \frac{\delta e_T}{\delta l}$$

$$= \delta e_T \, \delta \dot{m}.$$

This, when summed over the whole control volume, gives the rate of change of total energy content of the control volume

E_T, as

$$\frac{DE_T}{Dt} = \int_v de_T \, d\dot{m}$$

$$= \int d\dot{m} \int de_T$$

$$= \int (e_{T\,\text{out}} - e_{T\,\text{in}}) \, d\dot{m}$$

$$= \int_{\text{out}} e_T \, d\dot{m} - \int_{\text{in}} e_T \, d\dot{m},$$

which are the fifth and eighth terms in eqn. (15.24).

Using all these results together with eqn. (5.8) results in eqn. (15.24) being expressed in the form

$$\dot{W}_{\tau s\,\text{to}} + \dot{W}_{\sigma s\,\text{to}} + \dot{W}_{\beta\,\text{to}} + \dot{Q}_{s\,\text{to}} = \frac{DE_T}{Dt}$$

or
$$\dot{W}_{\text{to}} + \dot{Q}_{\text{to}} = \frac{DE_T}{Dt}.$$

In accordance with the assumptions and sign convention implicit in eqn. (6.9), this can be rewritten

$$\dot{Q} - \dot{W} = \frac{DE_T}{Dt}. \qquad (15.25)$$

It is seen that the rate at which work is being done to the control volume, plus the rate at which heat is being applied to it equals the rate at which the energy content of the control volume is changing.

This result applies, at any instant, to the system consisting of the fluid contained within a control volume and for any other succeeding control volume that the system may later occupy. Thus it applies to the process of a system and can then be integrated with respect to time between any two times, during the process t_1 and t_2, say, giving

$$\int_1^2 \dot{W}_{to}\,dt + \int_1^2 \dot{Q}_{to}\,dt = \int_1^2 \frac{DE_T}{Dt}\,dt,$$

which results in

$$\Big|_1^2 W_{to} + \Big|_1^2 Q_{to} = E_{T2} - E_{T1}$$

or, consistent with eqn. (15.25),

$$\Big|_1^2 Q - \Big|_1^2 W = E_{T2} - E_{T1}.$$

15.11 The energy relation for flow with diffusion

In incorporating the effects of diffusion into the energy equations just derived for a steady flow, a previously used technique is adopted. In § 8.13 the effects of mechanical and thermodynamic processes were combined by considering a series of small processes which were alternately mechanical and thermodynamic. Now a diffusion process is added.

An additive property of a homogeneous substance can be divided evenly amongst the particles forming that substance. Then when diffusion takes place the diffusion flow carries these properties with it. For instance, the internal energy can be divided uniformly amongst the particles[†] and then the internal energy is carried with the particles as they diffuse. Thus internal energy can be transported into a control volume so that the initial energy content of the control volume can be changed before the start of a subsequent flow process. Similarly, the

[†] In fact the energies of the particles cover a range of values, but this even division is acceptable here if we are considering a large number of particles.

additive property p/ϱ can be transported into a control volume thereby enhancing the ability of the fluid there to do work upon its surroundings.

Rearranging eqn. (15.11) gives

$$\frac{De}{Dt}\delta t + \delta\left(\frac{p}{\varrho}\right) = -\frac{D}{Dt}\left(\frac{1}{2}q^2\right)\delta t - g\,\delta z + \frac{\dot{W}_{\tau \text{to}}}{\delta \dot{m}} + \frac{\dot{Q}_{\text{to}}}{\delta \dot{m}}. \tag{15.26}$$

The left-hand side of this equation is the change in $(e+p/\varrho)$ in a small process occurring in a time δt. Or, from eqn. (15.19)

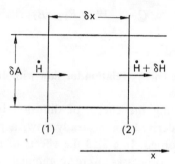

Fig. 15.3

it is the change in the enthalpy h. If now one considers a diffusion process, for which there is no rate of mass flow, then the enthalpy is the only additive property the transport of which can affect the starting conditions for a solution of eqn. (15.26).

To illustrate the transport of enthalpy, Fig. 15.3 is a diagram of a straight streamtube of cross-sectional area δA and which is aligned in the x-direction. Two planes denoted (1) and (2) in this figure are drawn to enclose a control volume of length δx containing a mass δm.

From eqn. (12.23) the rate at which particles are crossing section (1) is

$$\frac{N}{\delta t} = (c_1 v_1 + c_2 v_2)\frac{\delta A}{A_m}.$$

Consistently with eqn. (5.4), a total enthalpy H can be defined by

$$H \equiv h\, \delta m.$$

If $\bar{h}$ is the mean enthalpy per particle,[†] then the rate at which the total enthalpy is being transported $\dot{H}$ is given by

$$\dot{H} = \sum \frac{N_i}{\delta t} \bar{h}$$
$$= (c_1 v_1 \bar{h}_1 + c_2 v_2 \bar{h}_2)\frac{\delta A}{A_m}. \qquad (15.27)$$

Now the enthalpy of each constituent alone is

$$h_i = \frac{N_i \bar{h}_i}{\delta m_i}$$
$$= \frac{N_i \bar{h}_i}{N_i M_i A_m}$$
$$= \frac{\bar{h}_i}{M_i A_m}.$$

Thus using this relation and eqn. (12.21), eqn. (15.27) becomes

$$\dot{H} = (\varrho_1 v_1 h_1 + \varrho_2 v_2 h_2)\,\delta A.$$

It has already been stated that the diffusion process being considered is one in which there is no net rate of mass flow. Thus $\dot{x}$ as defined by eqn. (12.29) is zero and so the last equation can be written

$$\dot{H} = [\varrho_1 h_1 (v_1 - \dot{x}) + \varrho_2 h_2 (v_2 - \dot{x})]\,\delta A$$
$$= \sum \varrho_i h_i (v_i - \dot{x})\,\delta A. \qquad (15.28)$$

The rate of diffusion of enthalpy across section (2) that is

[†] It is important to observe that $\bar{h}$ may not be a simple additive function of the proportions of a mixture.[3]

shown in Fig. 15.3 can be written

$$\dot{H} + \delta\dot{H} = \dot{H} + \frac{\partial \dot{H}}{\partial x} \delta x.$$

Thus the net rate at which enthalpy is flowing into the elemental control volume $\dot{H}_{to}$ is given by

$$\dot{H}_{to} = \dot{H} - \left(\dot{H} + \frac{\partial \dot{H}}{\partial x} \delta x\right)$$

$$= -\frac{\partial \dot{H}}{\partial x} \delta x,$$

and from eqn. (15.28) this is

$$\dot{H}_{to} = -\delta x \frac{\partial}{\partial x} \sum \varrho_i h_i (v_i - \dot{x}) \, \delta A$$

$$= -\delta x \, \delta A \sum \frac{\partial}{\partial x} [\varrho_i h_i (v_i - \dot{x})]. \qquad (15.29)$$

In a time δt the transport of total enthalpy into the control volume is

$$\dot{H}_{to} \, \delta t$$

so that the transport of enthalpy is

$$\frac{\dot{H}_{to} \, \delta t}{\delta m}.$$

Making the time of this diffusion process the same as that of the subsequent flow process, then this transport becomes

$$\frac{\dot{H}_{to}}{\delta \dot{m}}.$$

In this diffusion process, as given by eqn. (12.29) $\dot{x} = 0$. Thus a flow process which follows the diffusion process must occur by the addition of a velocity relative to this value. The addition of this flow velocity to both v_1 and v_2 results, from eqn. (12.29), in a similar addition to $\dot{x}$ and hence from eqn. (15.29)

LAW OF THERMODYNAMICS TO FLUIDS IN MOTION

leaves the diffusion process unaffected because the latter is controlled by a difference in velocities.

Adding the changes in enthalpy given by eqns. (15.26) and (15.29) results in

$$\frac{De}{Dt}\delta t + \delta\left(\frac{p}{\varrho}\right) = -\frac{D}{Dt}\left(\frac{1}{2}\dot{x}^2\right)\delta t - g\,\delta z + \frac{\dot{W}_{\tau\text{to}}}{\delta \dot{m}} + \frac{\dot{Q}_{\text{to}}}{\delta \dot{m}} + \frac{\dot{H}_{\text{to}}}{\delta \dot{m}}. \qquad (15.30)$$

This relation can also be derived using the kinetic theory of gases.[4] As already described, if now diffusion and flow processes succeed one another in succession, in the limit as the time of each process becomes infinitesimally small then there results a smooth, combined process at the rates given by the above equation.

The summation of the enthalpy transport for a succession of elements follows the lines already laid down for the summation of the work and heat terms. Thus in analogy with eqn. (15.12), eqn. (15.30) leads to

$$\frac{p_1}{\varrho_1} - \frac{p_2}{\varrho_2} + gz_1 - gz_2 + \frac{1}{\delta \dot{m}}\sum_1^2 \dot{W}_{\tau\text{to}} + \frac{1}{\delta \dot{m}}\sum_1^2 \dot{Q}_{\text{to}} + \frac{1}{\delta \dot{m}}\sum_1^2 \dot{H}_{\text{to}} = e_{T2} - e_{T1} \qquad (15.31)$$

and, analogous with eqn. (15.24), the control volume equation, becomes

$$\dot{W}_{\tau s\,\text{to}} + \dot{Q}_{s\,\text{to}} + \dot{H}_{s\,\text{to}} = \int_{\text{out}}\left(\frac{p}{\varrho} + gz + e_T\right)d\dot{m} - \int_{\text{in}}\left(\frac{p}{\varrho} + gz + e_T\right)d\dot{m}. \qquad (15.32)$$

The enthalpy transport given by eqn. (15.29) can be written in terms of the diffusion coefficient through the following analysis.

From eqns. (12.24) and (12.26)
$$\dot{n}_1 = c_1(v_1 - v^*)$$
$$= c_1 v_1 - \frac{c_1}{c}(c_1 v_1 + c_2 v_2),$$

so that

$$v_2 = \frac{c}{c_1 c_2}\left[c_1 v_1 \left(1 - \frac{c_1}{c}\right) - \dot{n}_1\right].$$

Similarly, from eqn. (12.29)

$$v_2 = \frac{\varrho}{\varrho_1 \varrho_2}\left\{\varrho_1 v_1 \left(1 - \frac{\varrho_1}{\varrho}\right) - [\varrho_1(v_1 - \dot{x})]\right\}$$

Eliminating v_2 between these two equations and using eqn. 12.21 leads to

$$\varrho_1(v_1 - \dot{x}) = M_1 M_2 \frac{c}{\varrho} \dot{n}_1 \qquad (15.33)$$

From eqn. 12.33,

$$\dot{n}_1 = -cD \frac{\partial}{\partial x}\left[\frac{1}{1 + (c_2/c_1)}\right]$$
$$= \frac{cD}{[1 + (c_2/c_1)]^2}\left(\frac{1}{c_1}\frac{dc_2}{dx} - \frac{c_2}{c_1^2}\frac{dc_1}{dx}\right)$$
$$= \frac{D}{c}\left(c_1 \frac{dc_2}{dx} - c_2 \frac{dc_1}{dx}\right)$$
$$= \frac{D}{cM_1 M_2}\left(\varrho_1 \frac{d\varrho_2}{dx} - \varrho_2 \frac{d\varrho_1}{dx}\right). \qquad (15.34)$$

Now

$$\frac{d}{dx}\left(\frac{\varrho_1}{\varrho}\right) = \frac{d}{dx}\left[\frac{1}{1 + (\varrho_2/\varrho_1)}\right]$$
$$= -\frac{1}{\varrho^2}\left[\varrho_1 \frac{d\varrho_2}{dx} - \varrho_2 \frac{d\varrho_1}{dx}\right].$$

Substituting this and eqn. (15.34) into eqn. (15.33) gives

$$\varrho_1(v_1 - \dot{x}) = -\varrho D \frac{d}{dx}\left(\frac{\varrho_1}{\varrho}\right). \qquad (15.35)$$

Thus use of eqns. (15.29) and (15.35) gives

$$\frac{\dot{H}_{\text{to}}}{\delta \dot{m}} = \frac{\delta t}{\varrho} \sum \frac{\partial}{\partial x}\left[h_i \varrho D \frac{\partial}{\partial x}\left(\frac{\varrho_i}{\varrho}\right)\right]$$

$$= \delta t \frac{D}{\varrho} \sum \frac{\partial}{\partial x}\left[\bar{h}_i \varrho \frac{\partial}{\partial x}\left(\frac{\varrho_i}{\varrho}\right)\right].$$

Making δt the same for both infinitesimal processes then for the flow process $\delta x = \dot{x} \, \delta t$ and so

$$\frac{\dot{H}_{\text{to}}}{\delta \dot{m}} = \frac{D}{\varrho \dot{x}} \sum \frac{\partial}{\partial x}\left[h_i \varrho \frac{\partial}{\partial x}\left(\frac{\varrho_i}{\varrho}\right)\right] \delta x.$$

Inserting a reference length x_0, say, this can be written

$$\frac{\dot{H}_{\text{to}}}{\delta \dot{m}} = \frac{\mu}{\varrho x_0 \dot{x}} \frac{D\varrho}{\mu} \sum \frac{1}{\varrho} \frac{\partial}{\partial(x/x_0)}\left[h_i \varrho \frac{\partial}{\partial(x/x_0)}\left(\frac{\varrho_i}{\varrho}\right)\right] \delta\left(\frac{x}{x_0}\right). \qquad (15.36)$$

The quantity $\mu/(D\varrho)$ is called the Schmidt number. Examples of its numerical value are about 0·75 for mixtures of gases and about 10^7 for ions diffusing in hydrocarbon liquids. The quantity $\varrho x_0 \dot{x}/\mu$ is the Reynolds number. Thus for the term of eqn. (15.36) to be significant in gas flows at common values of the Reynolds number, the concentration gradients must be very high, and this usually only happens in the flow in thin boundary layers where mass is diffusing across the flow or in shock waves.[5]

In liquids, where the Schmidt and Reynolds numbers are often comparable, the term of eqn. (15.36) is then negligible.†

† There is always a danger in comparisons of this sort arising from the fact that, usually, interest is in changes in each term.

When this one-dimensional flow is incompressible, the summation of eqn. (15.36) along a streamline as indicated in eqn. (15.31) leads to

$$\frac{1}{\delta \dot{m}} \sum_1^2 \dot{H}_{to} = \frac{\mu}{\varrho x_0 \dot{x}} \frac{D\varrho}{\mu} \int_1^2 d\left[\sum h_i \frac{\partial}{\partial(x/x_0)}\left(\frac{\varrho_i}{\varrho}\right)\right]$$

$$= \frac{\mu}{\varrho x_0 \dot{x}} \frac{D\varrho}{\mu} \left\{\left[\sum h_i \frac{\partial}{\partial(x/x_0)}\left(\frac{\varrho_i}{\varrho}\right)\right]_2 - \left[\sum h_i \frac{\partial}{\partial(x/x_0)}\left(\frac{\varrho_i}{\varrho}\right)\right]_1\right\},$$

the suffixes 1 and 2 here and in eqn. (15.31) referring to two positions on the streamline.

References

1. LIEPMANN, H. W. and ROSHKO, A., *Elements of Gas Dynamics*, Wiley, New York, 1957, § 13.2, p. 306.
2. TAYLOR, G. I. and MACCOLL, J. W., *The Mechanics of Compressible Fluids*, H I 6, p. 218, *Aerodynamic Theory* (Ed. W. F. Durand), vol. 3, Springer, Berlin, 1934.
3. MONTGOMERY, S. R., *Second Law of Thermodynamics*, Pergamon, Oxford, 1966, p. 100.
4. HIRSCHFELDER, J. O., CURTISS, C. F., and BIRD, R. B., *Molecular Theory of Gases and Liquids*, Wiley, New York, 1954, p. 698.
5. COWLING, T. G., The influence of diffusion on the propagation of shock waves, *Phil. Mag.* **33** (7th Ser), 61 (Jan. 1942).

CHAPTER 16

THE ADIABATIC FLOW

16.1 The non-viscous flow along a streamline

In a flow along a streamline in which the effects of viscosity, of heat conduction and of diffusion are negligible, the momentum equation (14.2) can be rewritten as

$$\frac{dp}{\varrho}+g\,dz+d\left(\frac{1}{2}q^2\right) = 0, \qquad (16.1)$$

and the energy equation, (15.11), has the form

$$d\left(\frac{p}{\varrho}\right)+g\,dz+d\left(\frac{1}{2}q^2\right)+du = 0. \qquad (16.2)$$

In the case of an incompressible flow, comparison of these equations gives

$$du = 0 \quad \text{or} \quad u = \text{constant}.$$

Thus the flow of a liquid that satisfies eqn. (9.25) and has a constant value of C_u obeys the relation

$$T = \text{constant}; \qquad (16.3)$$

there is no change in temperature along a streamline.

In the case of the compressible flow of a gas satisfying eqn. (9.30) and again having a constant value of C_u, comparison of eqns. (16.1) and (16.2) leads to

$$d\left(\frac{p}{\varrho}\right)+C_u\,dT = \frac{dp}{\varrho}.$$

If the gas also satisfies the equation of state, eqn. (9.11), then

$$d\left(\frac{p}{\varrho}\right) + \frac{C_u M}{R} d\left(\frac{p}{\varrho}\right) - \frac{dp}{\varrho} = 0.$$

Forming the differential of p/ϱ gives

$$\left(1 + \frac{C_u M}{R}\right)\left(\frac{dp}{\varrho} - \frac{p}{\varrho^2} d\varrho\right) - \frac{dp}{\varrho} = 0$$

and so

$$\frac{dp}{p} - \left(\frac{R}{MC_u} + 1\right) \frac{d\varrho}{\varrho} = 0.$$

From eqn. (15.21),

$$\frac{R}{MC_u} + 1 = \frac{C_h}{C_u},$$

and this ratio is given the symbol γ; that is

$$\gamma \equiv \frac{C_h}{C_u}. \tag{16.4}$$

The above equation is now

$$\frac{dp}{p} - \gamma \frac{d\varrho}{\varrho} = 0,$$

and integration gives

$$\log p - \gamma \log \varrho = \text{constant}$$

and so that

$$p/\varrho^\gamma = \text{constant}. \tag{16.5}$$

This important relation is thus seen to govern the process occurring in the flow of an ideal gas along a streamline when viscous, heat conduction, and diffusion effects are negligible. The existence of the two equations (9.11) and (16.5), that relate the properties of the gas, means that only one property is needed to define the state during such a process: the gas becomes a one property substance and this fact can greatly aid the analysis of a flow.[1]

16.2 Stagnation properties

If, in a flow that is governed by eqn. (16.5), the fluid is brought to rest at some point on the streamline, the properties there are called the stagnation properties, and the point is called the stagnation point. Examples are the stagnation pressure[†] p_s, the stagnation density ϱ_s, and the stagnation temperature T_s. These properties are useful reference values in a flow. For instance, eqn. (15.23) can be written

$$C_h T + \tfrac{1}{2} q^2 = \text{constant},$$

and as $T = T_s$, when $q = 0$, this can be written

$$C_h T + \tfrac{1}{2} q^2 = C_h T_s. \tag{16.6}$$

Or, if a stagnation enthalpy h_s is used, this equation, together with eqn. (15.22) for the idealized gas, gives

$$h + \tfrac{1}{2} q^2 = h_s.$$

Similarly, eqn. (16.5) can be written

$$p/\varrho^\gamma = p_s/\varrho_s^\gamma. \tag{16.7}$$

It is important to remember here that in general these relations do not apply to flows in which there are viscous effects. For example, in the flow in a boundary layer with no heat being applied to the fluid at the wall, the temperature at the wall is neither the temperature nor the stagnation temperature of the flow in the mainstream, and the difference between the wall temperature and this stagnation temperature increases with increase of the stream velocity.[2]

It is useful that eqns. (16.3) and (16.5) can be shown to apply also to an unsteady flow.[3]

[†] This symbol is used here to make clear the distinction between this compressible flow and the incompressible flow discussed in § 14.1.

16.3 The compressible flow Bernoulli equation

For the process governed by eqn. (16.5), the integral occurring in the momentum equation, eqn. (14.3), can now be evaluated. It becomes

$$\int \frac{dp}{\varrho} = \frac{p_s^{\frac{1}{\gamma}}}{\varrho_s} \int \frac{dp}{p^{\frac{1}{\gamma}}}$$

$$= \frac{p_s^{\frac{1}{\gamma}}}{\varrho_s} \frac{p^{1-\frac{1}{\gamma}}}{1-(1/\gamma)} + \text{constant}$$

$$= \frac{\gamma}{\gamma-1} \frac{p_s}{\varrho_s} \left(\frac{p}{p_s}\right)^{\frac{\gamma-1}{\gamma}} + \text{constant}.$$

Then eqn. (14.3) becomes

$$\frac{\gamma}{\gamma-1} \frac{p_s}{\varrho_s} \left(\frac{p}{p_s}\right)^{\frac{\gamma-1}{\gamma}} + gz + \frac{1}{2} q^2 = \text{constant}$$

and, arbitrarily putting $z = 0$ at the stagnation point, enables this to be written

$$\frac{\gamma}{\gamma-1} \frac{p_s}{\varrho_s} \left(\frac{p}{p_s}\right)^{\frac{\gamma-1}{\gamma}} + gz + \frac{1}{2} q^2 = \frac{\gamma}{\gamma-1} \frac{p_s}{\varrho_s}, \quad (16.8)$$

which is the compressible flow version of the Bernoulli equation.

16.4 Flow with stationary boundaries

An interesting application of the energy equation for a control volume, eqn. (15.24), is to the flow through a duct such as that sketched in Fig. 16.1. This flow is shown as being from a region at station 1 to another at station 2. Many such real

THE ADIABATIC FLOW

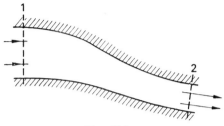

Fig. 16.1

flows will satisfy the following conditions which result in the consequences to be described.

(a) Either the walls between stations 1 and 2 can be effectively heat insulating ones having a value of the thermal conductivity k that is much lower than that of the fluid, or the walls can be maintained at a temperature that makes the gradient of the temperature, in the fluid and normal to the wall, zero. This means that the contribution to the $\dot{Q}_{s\,\text{to}}$ term by heat applied at the walls is effectively zero.

(b) The streamwise temperature gradients at 1 and 2 can be moderate so that, as discussed in § 15.7, the contribution here to $\dot{Q}_{s\,\text{to}}$ is also zero.

(c) All parts of the boundary are stationary, and so the boundary condition for a real flow being that of zero velocity, there will be no contribution to $\dot{W}_{\tau s\,\text{to}}$ along the duct walls.

(d) The stations 1 and 2 can be so drawn that the fluid velocity is everywhere perpendicular to them. Thus again there will be no contribution there to $\dot{W}_{\tau s\,\text{to}}$.

As a result of the conditions (a) and (b), in the energy relation for a control volume

$$\dot{Q}_{s\,\text{to}} = 0.$$

This type of flow is called an adiabatic one: the non-heat conducting flow discussed in § 16.1 is something more than this.

As a result of conditions (c) and (d), in the energy equation for a control volume

$$\dot{W}_{\tau s \text{ to}} = 0,$$

even though viscous effects within the flow may be quite marked.

16.5 Volume change of a gas

It now becomes possible to reconsider the process of the adiabatic compression of a gas within a cylinder by the movement of a piston. This is a flow process for the gas between two states, each being of uniformly zero velocity. The flow cannot be regarded as a steady one because, for instance, eqns. (16.6), (16.8), and (16.7) give, respectively, for the change between state A and state B,

$$T_A = T_B,$$
$$p_A = p_B,$$
$$\varrho_A = \varrho_B,$$

and the last of these is not true because the density increases as the volume of the fixed mass of gas that is enclosed in the cylinder is reduced by a compressive movement of the piston.

The work done to the gas by the piston can now be calculated. From eqn. (6.3)

$$\left|_A^B W_{\text{to}} = (u_B - u_A)m,\right.$$

where m is the mass of the gas. Using eqns. (9.31), (9.11), and (15.21),

$$u = \frac{1}{\gamma - 1} \frac{p}{\varrho}, \qquad (16.9)$$

THE ADIABATIC FLOW

so that remembering that the end states are motionless ones where the pressure and density will be single valued,

$$\left|_A^B W_{to} = \frac{m}{\gamma-1}\left(\frac{p_B}{\varrho_B} - \frac{p_A}{\varrho_A}\right).$$

For the particular type of unsteady flow governed by eqn. (16.5),

$$\frac{p_B}{\varrho_B} = \frac{p_A}{\varrho_A}\left(\frac{\varrho_B}{\varrho_A}\right)^{\gamma-1}.$$

Thus

$$\left|_A^B W_{to} = \frac{m}{\gamma-1}\frac{p_A}{\varrho_A}\left[\left(\frac{\varrho_B}{\varrho_A}\right)^{\gamma-1} - 1\right].$$

For a fixed mass of gas m the density will be inversely proportional to the volume of the gas V. Thus,

$$\frac{\varrho_B}{\varrho_A} = \frac{V_A}{V_B} \equiv r,$$

where r is called the compression ratio. Thus, finally,

$$\left|_A^B W_{to} = \frac{m}{\gamma-1}\frac{p_A}{\varrho_A}(r^{\gamma-1} - 1). \qquad (16.10)$$

It might be thought that the combined use of eqns. (8.5) and (16.5) might provide an alternative approach to this problem. However, the reasons for the reservations to this are now apparent. The use of eqn. (16.5) requires a high Reynolds number of the flow to justify the neglect of viscous effects, and this implies high velocities and consequent variations of velocity. Then from eqn. (16.8) the higher the velocity variations the greater do the pressures differ from uniformity. Thus eqn. (8.5), which assumes a uniform pressure, becomes reduced in accuracy as the viscous effects are reduced. But, indeed, eqn. (8.5) is not needed as the derivation of eqn. (16.10) shows.

16.6 The energy equation for fluid machines

The energy relation for a control volume, eqn. (15.24), was derived from that for a streamline, eqn. (15.12). The latter shows that for the non-viscous, non-heat conducting, type of flow described in § 16.1, the stagnation enthalpy is a constant along the flow. But a machine is put into a flow to change the stag-

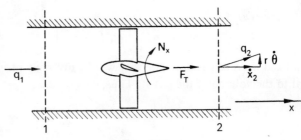

Fig. 16.2

nation enthalpy and so, as in the case of the momentum equation, an unsteady flow term must be introduced into the energy equation to account for this.

This is illustrated by a reconsideration of the discussion in § 14.9 on the flow of an incompressible fluid as it is pumped tahroug da uct by a fan. This flow is sketched in Fig. 16.2.

As a result of the rotation of the fan at an angular velocity ω, a thrust F_T, and a torque N_x is applied to the fluid contained between the duct walls and the two stations indicated 1 and 2 in this figure. To simplify discussion it is assumed that upstream the flow is uniformly in an axial direction at a velocity q_1, and that downstream the velocity has a uniform component $\dot{x}_2$ and a circumferential component $r\dot{\theta}$ impressed upon the flow by the torque N_x.

THE ADIABATIC FLOW

Thus, from the continuity equation, (12.11),

$$q_1 = \dot{x}_2. \tag{16.11}$$

From the momentum equation, (14.26),

$$p_1 - p_2 + \frac{F_T}{a} = 0, \tag{16.12}$$

and from the angular momentum equation, (14.22),[†]

$$N_x = \int_2 r^2 \dot{\theta}\, dm. \tag{16.13}$$

If the fan is rotating steadily and is regarded as an adiabatic mechanical system, then the rate at which it is doing work to the fluid is equal to the rate at which work is being done to drive it and this is ωN_x. The total rate at which work is being done to a control volume enclosing only the fluid is then

$$\dot{m}\left[\frac{p_1}{\varrho} - \frac{p_2}{\varrho}\right] + \omega N_x.$$

If the flow is an adiabatic one so that $\dot{Q}_{sto} = 0$, then, assuming the z-terms to be negligible, the energy relation, eqn. (15.24), becomes

$$\frac{p_1}{\varrho} - \frac{p_2}{\varrho} + \frac{\omega N_x}{\dot{m}} = u_2 - u_1 + \frac{1}{\dot{m}}\int \frac{1}{2} q_2^2\, dm \\ - \frac{1}{\dot{m}}\int \frac{1}{2} q_1^2\, dm + (UT), \tag{16.14}$$

where (UT) denotes the terms due to the unsteadiness in the flow.

Now downstream,

$$q_2^2 = \dot{x}_2^2 + r\dot{\theta}^2$$

and so, noting eqn. (16.11),

$$\frac{1}{2\dot{m}}\int (q_2^2 - q_1^2)\, dm = \frac{1}{2\dot{m}}\int (r\dot{\theta})^2\, dm,$$

[†] See footnote, p. 253.

and thus eqn. (16.14) becomes

$$\frac{p_1}{\varrho} - \frac{p_2}{\varrho} + \frac{\omega N_x}{\dot{m}} = u_2 - u_1 + \frac{1}{2\dot{m}} \int (r\dot{\theta})^2 \, d\dot{m} + (UT). \quad (16.15)$$

As in the discussion in § 14.9, an angular velocity of ω is applied to the whole system to bring the fan to rest and make the flow a steady one. As previously demonstrated in that section, this has no effect upon the value of (p_1-p_2). It is shown in § 16.1 that in this incompressible adiabatic flow the value of (u_2-u_1) is due solely to viscous effects. As discussed in § 14.9, addition of the angular rotation to the whole flow does not alter the viscous effects and so does not change the value of (u_2-u_1).

The circumferential velocity at the downstream station is now $(r\omega - r\dot{\theta})$, and so q^2 downstream has the value

$$\dot{x}_2^2 + (r\omega - r\dot{\theta})^2$$

and upstream it has the value,

$$\dot{x}_1^2 + r^2\omega^2.$$

Thus the energy equation for this case when the fan blades are stationary is

$$\frac{p_1}{\varrho} - \frac{p_2}{\varrho} = u_2 - u_1 + \frac{1}{2\dot{m}} \int \{[\dot{x}_2^2 + r^2(\omega-\dot{\theta})^2] - [\dot{x}_1^2 + r^2\omega^2]\} \, d\dot{m}.$$

As the continuity equation now gives

$$\dot{x}_1 = \dot{x}_2,$$

this equation becomes

$$\frac{p_1}{\varrho} - \frac{p_2}{\varrho} = u_2 - u_1 + \frac{1}{2\dot{m}} \int (r^2\dot{\theta}^2 - 2r^2\omega\dot{\theta}) \, d\dot{m}.$$

Subtracting this from eqn. (16.15) gives

$$(UT) = -\frac{1}{\dot{m}} \int r^2\omega\dot{\theta} \, d\dot{m} + \frac{\omega N_x}{\dot{m}}$$
$$= \frac{\omega}{\dot{m}} \int r^2\dot{\theta} \, d\dot{m} + \frac{\omega N_x}{\dot{m}},$$

THE ADIABATIC FLOW

and substituting from eqn. (16.13) gives
$$(UT) = 0.$$
The energy equation, (16.15), thus finally becomes
$$\frac{p_1}{\varrho} - \frac{p_2}{\varrho} + \frac{\omega N_x}{\dot{m}} = u_2 - u_1 + \frac{1}{2\dot{m}} \int (r\theta)^2 \, d\dot{m}.$$

The term $(\omega N_x)/\dot{m}$ is the work that is being applied to the fan by its motor per unit mass of fluid flowing through the duct. It seems a general result for fluid machines of various types, even in compressible flow, that the unsteady flow term in the control volume energy equation is zero.

Substituting from eqn. (16.12) gives
$$\frac{\omega N_x}{\dot{m}} - \frac{F_T q_1}{\dot{m}} = u_2 - u_1 + \frac{1}{2\dot{m}} \int (r\theta)^2 \, d\dot{m}.$$

Because the downstream flow has a rotating component, a circumferential shear force will act at the duct walls. Thus θ will gradually decrease downstream and when it becomes zero,
$$\frac{\omega N_x}{\dot{m}} - \frac{F_T q_1}{\dot{m}} = u_2 - u_1$$
which, for fluids obeying eqn. (9.29), becomes
$$\frac{\omega N_x}{\dot{m}} - \frac{F_T q_1}{\dot{m}} = C_u(T_2 - T_1). \qquad (16.16)$$

If, in the steady flow arrangement, the flow is free of viscous effects, then from eqn. (16.3)
$$T_1 = T_2,$$
and so eqn. (16.16) reduces to,
$$\frac{F_T q_1}{\dot{m}} = \frac{\omega N_x}{\dot{m}}. \qquad (16.17)$$

In a real flow viscous effects are present and it is found that
$$T_2 > T_1.$$

This means that

$$\frac{F_T q_1}{\dot{m}} < \frac{\omega N_x}{\dot{m}},$$

suggesting that the ideal value of the thrust is that given by eqn. (16.17). Hence a fan efficiency η is defined by

$$\eta \equiv \left(\frac{F_T q_1}{\dot{m}}\right) \bigg/ \left(\frac{\omega N_x}{\dot{m}}\right)$$

$$= \frac{F_T q_1}{\omega N_x}.$$

Substitution of this definition into eqn. (16.16) results in

$$(1-\eta)\frac{\omega N_x}{\dot{m}} = C_u(T_2 - T_1),$$

which forms the basis of a means of measuring efficiencies.[4]

16.7 The use of mean quantities

The definition of a mean velocity at a section across a flow has been described in § 12.4. The care with which this mean value must be used is made apparent by reference to the momentum and energy equations.

Postulating a section of flow forming the only part of a control volume across which the properties of the fluid and the flow are unknown, enables these three equations to be written as

$$\int d\dot{m} = A_1, \qquad (16.18)$$

$$\int p\, dA + \int q\, d\dot{m} = A_2, \qquad (16.19)$$

$$\int \frac{p}{\varrho}\, d\dot{m} + \int u\, d\dot{m} + \int \frac{1}{2} q^2\, d\dot{m} = A_3. \qquad (16.20)$$

The quantities A_1, A_2, and A_3 would be known through a specification of the rest of the control volume. The terms in z

THE ADIABATIC FLOW

are excluded as for a gas flow and the flow is assumed to be incompressible only to simplify discussion.

Noting eqn. (12.5), then the velocity terms in these three equations are, respectively,

$$\varrho \int q \, dA, \quad \varrho \int q^2 \, dA, \quad \text{and} \quad \frac{\varrho}{2} \int q^3 \, dA,$$

so that if, as in § 12.4, the first of these is used to define a mean velocity $\bar{q}$ it is not then generally permissible to write either

$$\int q^2 \, dA = (\bar{q})^2 A \qquad (16.21)$$
or
$$\int q^3 \, dA = (\bar{q})^3 A.$$

If, for instance, eqn. (16.21) is substituted into eqn. (16.19), then a mean pressure could be defined by

$$\bar{p} A = A_2 - (\bar{q})^2 A,$$

but it might not be equal to an actual pressure which was uniform.

16.8 Inadequacy of the governing equations

Simplifying the flow discussed in the last section to one of uniform flow of an ideal gas in the x-direction at the station where conditions are unknown, enables eqns. (16.18), (16.19), and (16.20) to be written respectively

$$\varrho \dot{x} A = A_1,$$
$$pA + \varrho \dot{x}^2 A = A_2,$$
$$p\dot{x}A + \varrho u \dot{x} A + \tfrac{1}{2} \varrho \dot{x}^3 A = A_3.$$

The unknowns are p, ϱ, $\dot{x}$, and u, that is four in number. Another equation is required and it is given by eqn. (16.9), that is

$$\frac{p}{\varrho} = (\gamma - 1) u.$$

From these four equations p, ϱ, and u can be eliminated leaving a single relation for $\dot{x}$. The result is

$$A_1 \frac{\gamma+1}{2(\gamma-1)} \dot{x}^2 - A_2 \frac{\gamma}{\gamma-1} \dot{x} + A_3 = 0.$$

This is a quadratic in $\dot{x}$ which thus has two solutions both of which can be real. The governing equations are sufficient to give a solution but are inadequate for deciding which solution gives the flow that occurs in reality. This can only be determined as a result of the discussion in Montgomery's *Second Law of Thermodynamics* (Pergamon, 1966) in the same series. Before passing to this, encouragement can be derived from the following conversation.[5]

"Fan her head!" the Red Queen anxiously interrupted. "She'll be feverish after so much thinking."

References

1. CHEERS, F., *Elements of Compressible Flow*, Wiley, London, 1963, p. 46.
2. SHAPIRO, A. H., *The Dynamics and Thermodynamics of Compressible Fluid Flow*, Ronald, New York, 1953, vol. 2, p. 1056.
3. RAYLEIGH, J. W. S., *The Theory of Sound*, 2nd edn., Macmillan, London, 1894, vol. 2, § 246, p. 21.
4. FOORD, T. R., LANGLANDS, R. C., and EL-AGIB, A. A. R., *New Developments in the Thermometric Method of Measuring Hydraulic Efficiency and Flow*, NEL Rep. 130, Jan. 1964.
5. CARROLL, L., *Through the Looking Glass and what Alice found there*, Macmillan, London, 1965, ch. 9, p. 202.

INDEX

Acceleration
　change as a result of force　6
　drag due to　251
　due to gravity　7
　due to viscous stresses　195
　giving delay effects　185
　normal to path　5
　of a fluid element　179
　polar coordinates　5
　relative to origin　5
　tangential to path　5
　x, y, components　5
Additive property　50
　of homogeneous system　52
Adiabatic flow　281, 286
Alice, difficulty with typical lecturer　60
Ampere, unit　38
Angstrom, unit　35
Angular momentum, control volume　251
Angular velocity　180, 191, 194
Archimedes principle　158, 187
Atmosphere
　imbalance of heat to　149, 158
　lack of thermal equilibrium　31
　temperature variation in　148
　unit of pressure　141
Atomic mass, definition (atomic weight)　34
Avogadro's constant　34
Axes, fixed and moving　176

Bernoulli equation　233
　for compressible flow　284
　variation of constant　237

Black body
　maximum emission　90
　radiation　88
Boundary condition
　streamline　174
　with diffusion　221
Boundary layer flow　200
　laminar　201
　separation　202
　transition　201
　turbulent　201
British thermal unit　35, 60

Calorie, unit　35, 59
Candela, unit　38
Celsius, temperature scale　37
Centigrade *see* Celsius
Centre of gravity　12
Centre of pressure　146
Centroid, centre of gravity　146
Coefficient of
　diffusion　222
　enthalpy　267
　friction　242
　internal energy　115
　linear thermal expansion　108
　　for rubber　113
　specific heat　115, 150, 267
　thermal expansion of liquid　129
　volume thermal expansion　108
Compressibility factor　136
Compressibility of a liquid　129
Compressible flow　202
Concentration
　molar　216
　velocity　217

Conduction of heat 79
Conservative and stationary field, height 2
Conservation of mass in a flow 204
Contact angle 162
Continuum 14
Continuity equation 210
Control volume 207
  conservation of mass for 209
    incompressible flow 211
    steady flow 211
  momentum equation for 243
Convection velocity 221
Critical point 171
Cycle
  cyclic process 67
  definition 11
  work done in by gravity 11

Delay effects in a flow 185
Density
  continuum limitation 17
  definition 16
  of a gas in motion 18
  partial 216
Differential, perfect 4
Diffusion
  coefficient of 222
  effect on continuity equation 221
    energy equation 273
    momentum equation 255
  rate 217
  self- 225
Discontinuity
  at phase boundary 168
  in height 3
Dissociation of a gas 139
Distance
  arbitrary origin 2
  path dependent 1
Drag forces, result of viscosity 248

Eddy, flow 197
Electrical charge, work by 73
Electrical film resistance thermometer 27
Electricity
  application to change internal energy 75
  computation of 76
  regarded as work 78
Energy
  equation for control volume 268
    flow 262
    machines 288
  equation with diffusion 273
  fusion 167
  internal 69
  kinetic
    definition 13
    a property 13
  of a field of force 74
  of photon concentration 87
  potential 75
  total 126, 262
  vaporization 167
Enthalpy 267
  coefficient of 267
Equation of state
  for gases 132
  for liquids 128
  for a metal 111
  for a rubber strip 113
  for a solid 110
  number of variables 112
  unaffected by speed of process 111
  use of pressure in a flow in 187
  Van der Waals' equation 137
  virial equation 137
Equilibrium state 48
  lack of in a flow 185
  zero work and heat 65
Extensive property *see* Additive property
Extinction coefficient of radiation 93

INDEX

Fahrenheit, temperature scales 38
Fan
  efficiency of 292
  pumping by 252
Fields of force 10
First law of thermodynamics 68
  application
    to rubber 121
    to solids 100
    to steel 119
  during a process 95
  rate equation
    in a fluid 270
    in a solid 95
Flat plate drag 248
Flow
  compressible, incompressible 202
  distortion in 180
  mean values in 292
  rates of 211
  rotation in 180
  steady, unsteady 174
  two-dimensional 180
  unsteady flow past aerofoil 248
  wake 246
Fluid
  describes liquids and gases 161
  distinction from solid 42
Force
  additive 207
  body forces 12
    causing rotation in a flow 191
    discontinuities 47
    proportional to mass 12, 47
  components 7
  definition 6
  on a surface immersed in a fluid 142, 144
  resulting acceleration change 6
  single valued function of time 8
  units of 34
  vector quantity 6
Free molecule flow 15
Friction
  coefficient of 242
  sliding process 100
Fusion energy 167

Gas
  non-isothermal 31
  two-property substance 139
Gas constant
  units conversion factor 140
  units of 134
  universal 133
Gaseous discharge tube 31
Gas thermometer 139
Governing equations 293
Gravity
  acceleration due to 7
  force 7
  standard value of g 36
  work done by on a solid 105

Head, total 233
Heat
  additive 62
  as a result of lack of thermal equilibrium 59
  conduction of 79
  continuity of gradient of rate 83
  directional nature of intensity of 96, 98
  evaluation at boundary 97
  in a bar 79
  manner of 79
  measure of 59
  no discontinuity in 82
  process in a solid 108
  rate intensity 64
  reaction of 63
  reversal of 63
  secondary measure of 61
  summation for system 64
Heat capacity *see* Internal energy coefficient

INDEX

Heat sink 66
Heat transfer *see* Heat, manner of
Height 2
H_2O (*see also* Ice, Water, Steam)
  ice point 37
  phases of 166
  triple point 37
Horizontal, definition 44
Hydrostatic pressure 141

Ice 166
Ice point of H_2O 37
Ideal gas 155
Incompressible flow 202
Infinitesimal element 257
Interaction between phenomena 229
Internal energy
  additive property 72
  a property 70
  definition 71
  non-differentiable 168
  of a field of force 74
    gas 152
    liquid 149
    metal 117
    mixture at a phase boundary 169
    rubber 118
    two-property substance 114
  per mass unit 72
  strain 121
  thermal 121
Internal energy coefficient 115, 150
  non-existent at phase boundary 168
  unsuitability as independent variable 154
Ionization of a gas 139
Irrotational flow 194
  related to Bernoulli constant 238
Isotherm 28

J, mechanical equivalent of heat 68
  values of 35, 69

Kelvin temperature 37
Kilogram, definition 34
Kinematic viscosity 197
Knudsen number 14

Laminar boundary layer 201
Length, units 34
Liquid drops, surface tension 166
Liquid
  compressibility of 129
  two-property substance 128

Macroscopic region 16
Manometer, surface tension 164
Mass
  absolute quantity 6
  conservation in a flow 204
    with diffusion 220
  definition 6
  flow rate of 207
  molar mass 34, 132
  units of 34
Mean values in a flow 292
Mechanical equivalent of heat, J 68, 69
  values of 35, 69
Meniscus 161
Mercury in glass thermometer 30
Metre, definition 34
Micron, definition 35
Microscopic region 16
Molar concentration 216
Molar mass 34, 132
Mole
  definition 34, 133
  volume of 133
Molecular mean path 14
Molecules, number of in gas 137

INDEX

Momentum
  definition 7
  equation for control volume 243
  equation for fluid motion 231
  equation with diffusion 255
  for incompressible flow 233
  unsteady flow terms 254
  varying mass 246

Natural convection 159
Newtonian fluids, definition 41
Newton's law
  statement of 6
  validity for control volume 246
Non-isothermal gas 31
Non-Newtonian fluids 43
Non-viscous flow 281
Number of molecules in gas 137

Onsager relations 230
Optical pyrometer 27

Partial pressures 140
Path
  distance dependent on 1
  height independent of 2
  of a process 49
Pathline
  coordinates 175
  definition 174
Phase boundaries 170
Phases 161
  changes in H_2O 166
  liquid–gas boundary 161
Photons 86
Piston, work done by pressures on 287
Poise, definition 184
Poisson's ratio, definition 106
Polhausen's parameter 241
Potential energy 75
Power, definition 9

Prandtl number 266
Pressure
  at a point in a fluid 22
  by molecular collisions 87
  cause of rotation 187
  components of force due to 155
  definition 22
  forces on fluid volumes 156
  horizontal constant 44
  hydrostatic on a solid 104
  in a flow 186
  non-existence in a flow 185
  photon collisions 87
  single-valued property in stationary gas 47
  total in a flow 233
    change of through a fan 252
  variation vertically
    in a gas 148
    in a liquid 141
  variation in a stationary fluid 44
Principle plane of stress 40
Process
  combination of mechanical and thermal 124
  cyclic 11, 67
  equations of phenomena 227
  of a system 39, 40
  rate of 94
  the stressing form of 102
Property
  additive 50, 51
  a characteristic 2, 25
  a quality 2
  arbitrary origin 49
  association of 52
  change independent of path 49
  definition 39
  development of 49
  distance 2
  fixing state by 26, 39
  height 2
  heterogeneous 50
  homogeneous 50

INDEX

Property (*cont.*)
  kinetic energy 13
  number to fix state 40
  one-property substance 282
  perfect differential 49
Pure substance *see* Two-property substance

Quality, definition 169
Queen, Red
  hazard of going to lecture of 60
  parting shot of 294

Radiation of heat 86
  absorption of 92
  extinction coefficient 93
  function of only temperature 88
  through an opaque medium 92
  view factor 90
Rankine temperature 38
Rate
  of a process 94
    of a fluid 270
    stretching of rod 122
Ratio of specific heats 282
Reversible process 104
  effect of Reynolds numbers 266
  non-existent at low speeds 266
Reynolds number 242, 266, 279
  of piston motion 287
Rotation in a flow 180, 187, 191
  zero in irrotational flow 194
Rupture of a liquid 41

Schmidt number 279
Second, unit 33
Self diffusion 225
Semi-permeable membrane 224
Separation of flow 202
  related to Polhausen parameter 241
Shear stress
  at a boundary in a flow 181
  in a turbulent flow 199
Shock wave 18, 279
Slip, lack of 101
Slip flow
  definition 15
  in a gas at a wall 43
Slug, definition 35, 36
Specific heat
  constant pressure *see* Enthalpy coefficient
  constant volume *see* Internal energy coefficient
  ratio of 282
Stagnation point 200
  pressure at 233
  properties at 283
Stagnation streamline 200
State (*see also* Equation of state)
  change of 40
  definition 26, 39
  of system in equilibrium 48
  one property substance 282
  plot of 48
Stationary and conservative field
  body forces 10
  height 4
Steady flow 174
  continuity equation for 211
  time derivatives zero 179
Steam 166
Stefan Law 88
Strain energy 121
Streamline 173
  stagnation 200
  variations perpendicular to 235
Streamtube 174
Stress
  continuity of 47
  definition 19
  in a solid 40
  normal in a fluid
    pressure 22
    principle 40
    sign of 20

INDEX

Stress (*cont.*)
  shear in a fluid   40, 181
    turbulent flow   199
Substance, pure *see* Two-property
Surface tension   19, 161
System
  contravention of
    by diffusion   215
    by electricity   76
  definition   39
  rate equation for   272

Temperature
  absolute scale   30, 139
  associated with thermal equilibrium   29
  at a wall   283
  change as a result of stretching a solid   120, 122
  continuum limitations   31
  definition   25, 29
  distribution in a bar   81
  no discontinuity in   80
  non-existence in a flow   185
  scales of   29, 37
  single-valued property   48
Thermal conductivity   80, 85
Thermal comparator   24
Thermal equilibrium   24, 28
Thermal expansion, coefficient of   108, 129
Thermal motion   159
Thermo-electric effects   230
Thermometer
  a thermal comparator   27
  characteristics   29
  gas   139
  numerical scale   30
  types of   27
Time
  derivatives zero   211
  increments positive   4
  lag effects in a flow   185
  single valued   9

  units of   33
Tonne, definition   35
Total
  energy   126
  head   233
  pressure   233, 252
Transition of boundary layer   201
Triple point   37, 171
Turbulence   197, 201
Two-property substance   114
  gas   139
  liquid   128

Units   33
Universal gas constant   133
Unsteady flow, validity of   283

Vaporization energy   167
Van der Waals' equation   137
Velocity
  concentration mean   217
  convection   221
  definition   4
  mean values in a flow   211
  profiles of boundary layer   201
  tangential to streamline   173
  zero at wall   41
Vertical, definition   44
View factor, radiation   90
Virial equation of state   137
Viscosity
  coefficient of   41
  effect of electric field   43
  kinematic   197
Viscous flow   238
Vortex flow   192
  shed by circular cylinder   197

Wake flow   200
  physical significance of   246
Wall condition in a flow   40
Water   166

INDEX

Weight, definition 7
Work
  additive nature 13, 207
  a happening during motion 8
  as a result of lack of mechanical equilibrium 59, 65
  by body forces 10, 55, 73
    hydrostatic pressure 104
    pressure 257, 287
    shear stresses in a flow 196, 242
    tensile force 102
  definition 7
  dependent on path 8, 119
  evaluation of 8, 56
  in stretching a rod 103
  not a differential 8
  on a piston 286
  reaction 10
  sign convention 7, 52
  usefulness of 53, 55

Young's modulus 103
  for rubber 113

Zero'th law of thermodynamics 26

to you who may —yu